DIVISOR THEORY IN MODULE CATEGORIES

NORTH-HOLLAND
MATHEMATICS STUDIES 14

Notas de Matemática (53)
Editor: Leopoldo Nachbin
Universidade Federal do Rio de Janeiro
and University of Rochester

Divisor Theory in Module Categories

W. V. VASCONCELOS

Rutgers University

1974

NORTH-HOLLAND PUBLISHING COMPANY - AMSTERDAM • OXFORD
AMERICAN ELSEVIER PUBLISHING COMPANY, INC. - NEW YORK

Library of Congress Catalog Card Number: 74-84871
North-Holland ISBN for this Series: 0 7204 2700 2
North-Holland ISBN for this Volume: 0 7204 2715 0
American Elsevier ISBN: 0 444 10737 1

PUBLISHERS:
NORTH-HOLLAND PUBLISHING COMPANY – AMSTERDAM
NORTH-HOLLAND PUBLISHING COMPANY, LTD. – OXFORD

SOLE DISTRIBUTORS FOR THE U.S.A. AND CANADA:
AMERICAN ELSEVIER PUBLISHING COMPANY, INC.
52 VANDERBILT AVENUE, NEW YORK, N.Y. 10017

PRINTED IN THE NETHERLANDS

Preface

Heuristically the divisor $\underline{d}(E)$ of an A-module E is the ideal of A packing the most information on E. A prime candidate for this role, the annihilator of E, lacks decent functorial properties. Instead, a generalization of another of the classical divisors plays a more visible role if one works in the following setting. Define a divisor on a subcategory $\mathcal{C}$ of mod(A) as an additive - with respect to short exact sequences - mapping from $\mathcal{C}$ into some semi-group $\mathcal{S}$ of ideals. An outstanding example is that found in the category $\mathcal{T}$ of finitely generated torsion modules of finite projective dimension over a Noetherian ring A. In this case one may define a divisor function from $\mathcal{T}$ into the semi-group Inv(A) of invertible ideals and obtain an exact sequence

$$\tilde{K}_0(A) \xrightarrow{\Delta} K_0(\mathcal{T}) \xrightarrow{\underline{d}} \mathrm{Inv}(A) \longrightarrow 1.$$

This $\underline{d}$ is defined in the usual manner : If

$$A^m \xrightarrow{\phi} A^n \longrightarrow E \longrightarrow 0$$

is a presentation of the module E, the determinantal ideal F(E) generated by the minors of order n of the matrix ϕ is changed into

$$\underline{d}(E) = (F(E)^{-1})^{-1}.$$

For the larger category of all finitely generated torsion modules this function is additive only if the ring is integrally closed. In fact, in this case, it is the only additive mapping satisfying $\underline{d}(A/xA) = xA$.

Divisors arise also in categories of modules generated by spherical modules - modules sharing certain homotopic proper -

ties of A or the canonical modules of Macaulay rings. Specifically we take a module G satisfying:

i) $\mathrm{Hom}_A(G,G) = A$, and

ii) $\mathrm{Ext}^i_A(G,G) = 0$ for $i > 0$

and consider modules susceptible of a representation

$$G^m \xrightarrow{\phi} G^n \longrightarrow E \longrightarrow 0.$$

As ϕ may be viewed as a matrix, an appropriate Schanuel's lemma will make the corresponding determinantal ideal well defined. As earlier the divisor will take values in Inv(A) whenever E admits a finite G-resolution. A consequence is that this divisor does not depend on the spherical module used in the finite resolution.

In the last chapter a divisor is defined in the category of modules of finite injective dimension but using the symmetrical notion of co-presentation.

These notes are a transcript of some lectures on divisor theory given at Rutgers University in the Spring of 1974. It was felt necessary to include an exposition of the homology of Noetherian - which was taken as synonymous with the theory of Macaulay rings - to make the whole sufficiently self-contained. Of course the time limits precluded any discussion of the interesting portion of that theory dealing with how Macaulay rings arise.

Major portions of these notes were jointly worked out with Jeffrey Dawson : this is at least the case for the whole of Chapter 5 and (3.18), the main result of Chapter 3. These and his ongoing research on higher divisorial ideals will also be part of his thesis. To Judith Sally we are indebted for valuable comments on an earlier version of the notes and corrections of some of the most offensive errors. Finally, the financial support of the National Science Foundation is gratefully acknowledged.

Table of Contents

Chapter 1

Local Algebra

In this chapter - of a sketchy nature - are grouped some of the basic facts of commutative algebra that shall be used throughout - Noetherian rings, Krull dimension, flatness, etc. [6, 20, 34] will be sources for the proofs, which will not, as a rule, be supplied here.

§1.1 Noetherian and coherent rings.

Although some of the notions discussed in these notes do not require the commutativity of the rings involved, this will be necessary for some constructions. Thus we assume the blanket condition that all rings are commutative, with 1, and that modules are unital.

Given a ring A and an A-module E we say that E is finitely generated if there is a surjection

$$A^n = \underbrace{A \oplus \dots \oplus A}_{n \text{ summands}} \longrightarrow E .$$

If $e_1,\dots,e_n$ are the images of the 'basis' elements of A^n, $v = (x_1,\dots,x_n) \in A^n$ will be called a relation on the e_i's if

$$\sum x_i e_i = 0 .$$

The set of such elements is a submodule K of A^n. E is said to be finitely presented if, besides, K is finitely generated. That the finiteness of K does not depend on how we present E follows from:

(1.1) Lemma. (Schanuel's lemma) Let

$$0 \longrightarrow K \longrightarrow P \longrightarrow E \longrightarrow 0 \quad \text{and}$$

$$0 \longrightarrow K' \longrightarrow P' \longrightarrow E \longrightarrow 0$$

be exact sequences with P and P' A-projectives. Then $K \oplus P' \cong K' \oplus P$.

Proof. Well-known.

Definitions. An A-module E is Noetherian if every submodule of E is finitely generated, or equivalently, if every ascending chain of submodules of E is stationary. As several constructions on modules show a finitary character, the following will be used at times. E is coherent if every finitely generated submodule is finitely presented. The ring A itself is Noetherian if it is Noetherian as an A-module, that is every ideal is finitely generated. Similarly for the notion of coherent ring.

(1.2) Lemma. If E is a finitely generated (resp. finitely presented) module over a Noetherian (resp. coherent) ring A then E is Noetherian (resp. coherent).

Proof. [6].

Remark. Although Noetherian rings are coherent this is no longer the case for modules as coherence is 'more' of a relative notion.

A chain of prime ideals (or primes for short) in A is a finite sequence

$$\underline{p}_0 \subset \underline{p}_1 \subset \cdots \subset \underline{p}_r$$

of distinct prime ideals of A. r is the length of the chain. The Krull dimension of A, K-dim(A), is the supremum of the lengths of all chains of prime ideals. The height of a prime $\underline{p}$, ht($\underline{p}$), is the supremum of the lengths of chains with elements contained in $\underline{p}$. If I is an ideal, ht(I) is defined as the infimum of the heights of primes containing I.

(1.3) Theorem. The height of a prime $\underline{p}$ in a Noetherian ring is finite and equal to the minimal number n of elements $x_1,\ldots,x_n \in \underline{p}$ such that $\underline{p}$ is a minimal prime ideal over $(x_1,\ldots,x_n)$.

Proof. [20, 34].

In particular, if A has a unique maximal ideal then K-dim(A) is finite. We shall refer to the elements $x_1,\ldots,x_n$ as a system of parameters for $\underline{p}$.

Given a commutative ring A and an A-module E, we say that a prime $\underline{p}$ is associated to E if $\underline{p}$ is minimal over an ideal of the form I = annihilator of $e \in E$. Ass(E) will denote the set of such primes. If $E \neq 0$, an element $x \in A$ is a zero-divisor for E if there is $0 \neq e \in E$ with $xe = 0$. The set of zero-divisors of E will be denoted z(E).

(1.4) Proposition. $z(E) = \bigcup_{\underline{p} \in \mathrm{Ass}(E)} \underline{p}$.

If A is Noetherian, each $\underline{p} \in \mathrm{Ass}(E)$ is actually the annihilator of an element of E. If, besides, E is finitely generated then Ass(E) is finite [20, 34].

(1.5) Proposition. If E is a finitely generated module over the Noetherian ring A there is a filtration

$$0 \subseteq M_1 \subseteq \dots \subseteq M_n = M,$$

such that $M_i/M_{i-1} \cong A/\underline{p}_i$, $\underline{p}_i$ = prime ideal.

Proof. [34].

§1.2 Local Rings.

The Jacobson radical of a ring A is the intersection J(A) of the maximal ideals of A. The ring A is *local* if it has a unique maximal ideal $\underline{m}$; $\underline{m}$ equals J(A) and is precisely the set of non-invertible elements of A. In this case $\underline{k} = A/\underline{m}$ will be called the residue field of A.

Let A be a local ring and E a finitely generated A-module. The minimum number of generators of E, $\nu(E)$, is the dimension of the $\underline{k}$-vector space $E/\underline{m}E$. Indeed, if $e_1, \dots, e_n$ is a set of elements of E whose images in $E/\underline{m}E$ form a basis, the submodule F of E they generate is such that $E = F + \underline{m}E$. That $E = F$ is a consequence of the ubiquitous

(1.6) Proposition. (Nakayama's lemma) Let E be a finitely generated A-module and F a submodule satisfying

$$E = F + J(A)\,E.$$

Then we have $F = E$.

Proof. If we let $G = E/F$, we have $G = JG$ and we

have to show $G = 0$. Let $g_1,\ldots,g_n$ be a system of generators of G. There exist elements

$$a_{ij} \in J \qquad 1 \le i,\ j \le n$$

such that

$$g_i = \sum_j a_{ij} g_j \ .$$

We have $\det(\delta_{ij} - a_{ij})G = 0$; since $\det(\delta_{ij} - a_{ij}) = 1 + a$, with $a \in J$, is invertible, $G = 0$.

Repeated use of (1.6) will be made in the following form: Let E be a finitely generated A-module and

$$\phi : E^m \longrightarrow E^n$$

a homomorphism realized by multiplication by a matrix (a_{ij}) $1 \le i \le m$, $1 \le j \le n$, with entries in J. If ϕ is surjective then $E = 0$.

A subset S of A is called a <u>multiplicative set</u> if $0 \notin S$ and $x,y \in S \Longrightarrow x \cdot y \in S$. If E is an A-module an equivalence relation is defined on $E \times S$ by $(e,x) \sim (f,y)$ iff there is $z \in S$ with $z(ye-xf) = 0$. $E \times S/\sim$ is written $S^{-1}E$ or E_S and has a natural group structure. Applied to the ring A itself it endows A_S with a ring structure and E_S is an A_S-module in the well known way. If $\underline{p}$ is a prime ideal, E_S for $S = A\setminus\underline{p}$ is called the localization of E at $\underline{p}$. As $\underline{p}$ itself is not a multiplicative set there will be no confusion when E_S is written $E_{\underline{p}}$. Viewed as a ring $A_{\underline{p}}$ has a simplified prime ideal structure:

its primes are in one-one correspondence with the primes of A contained in $\underline{p}$. In particular $A_{\underline{p}}$ is a local ring with maximal ideal $\underline{p}A_{\underline{p}}$. Its residue field will be denoted by $k(\underline{p})$.

The usefulness of this construction, that can be extended to module homomorphisms rests primarily on ([6]):

(1.7) Proposition.

i) If S is a multiplicative set and

$$E \xrightarrow{f} F \xrightarrow{g} G$$

is an exact sequence of modules, then

$$E_S \xrightarrow{f_S} F_S \xrightarrow{g_S} G_S$$

is also exact.

ii) $E = 0$ iff $E_{\underline{p}} = 0$ for every prime ideal.

A projective resolution of an A-module E is an exact sequence

$$P_{\bullet} : \ldots P_n \ldots P_1 \xrightarrow{f_1} P_0 \longrightarrow E \longrightarrow 0$$

with P_i A-projective. E is said to have projective dimension n ($\text{pd}_A E = n$) if there is a resolution with n least such that $P_{n+1} = 0$. In this case it follows immediately from (1.1) that whenever one starts building a resolution for E, the kernel of $P_{n-1} \xrightarrow{f_{n-1}} P_{n-2}$ is always projective. If E is a finitely presented module over a coherent ring then from (1.2) one concludes that E admits a projective resolution with finitely generated terms. In these cases a consequence of (1.7) is that $\text{pd}_A E = \sup \{\text{pd}_{A_{\underline{p}}} E_{\underline{p}}$, where $\underline{p}$ runs over the primes of A$\}$.

The <u>global dimension</u> of A is defined as $\sup\{ pd_A E,$ for all A-modules $\}$.

(1.8) <u>Theorem</u>. (Hilbert's syzygies theorem) If A is any ring

$$\text{gl dim}(A[t]) = \text{gl dim}(A) + 1,$$

where t is an indeterminate. In particular the polynomial ring in n indeterminates over a field has global dimension n.

An <u>injective resolution</u> of a module E is an exact sequence

$$I^{\bullet} : 0 \longrightarrow E \longrightarrow I^0 \longrightarrow I^1 \longrightarrow \dots \longrightarrow I^n \longrightarrow \dots$$

with I^i an injective A-module. The notion of injective dimen - sion is similarly defined while the 'global dimension ' of A coincides with that defined above.

Let A be a Noetherian local ring of maximal ideal $\underline{m}$ and residue field $\underline{k}$. Among the numbers

Krull dim(A),

$\dim_{\underline{k}}(\underline{m}/\underline{m}^2)$ and

global dim(A)

we have the following relations : gl dim(A) $\geq \dim_{\underline{k}}(\underline{m}/\underline{m}^2) \geq$ Krull dim(A) ([<u>32</u>]).

(1.9) <u>Theorem</u>. The following conditions are equivalent :

i) gl dim(A) $< \infty$.

ii) $\dim_{\underline{k}}(\underline{m}/\underline{m}^2)$ = Krull dim(A).

Moreover in either case one has equality of all three dimensions.

A ring satisfying these conditions will be called <u>regular</u>. The following gives a picture of regular rings ([<u>34</u>]):

(1.10) <u>Theorem</u>. Let A be a local ring. Then A is regular iff the graded ring associated to $\underline{m}$ is a polynomial ring.

Concretely, if $x_1,\ldots,x_n$ are the elements in a minimal generating set for $\underline{m}$ ($n = \dim_{\underline{k}}(\underline{m}/\underline{m}^2)$) we have a homomorphism of graded rings

$$\underline{h} : \underline{k}[X_1,\ldots,X_n] \longrightarrow k \oplus \underline{m}/\underline{m}^2 \oplus \ldots \oplus \underline{m}^r/\underline{m}^{r+1} \oplus \ldots$$

with $\underline{h}(X_i)$ = class of x_i in $\underline{m}/\underline{m}^2$. The assertion is that A is regular iff $\underline{h}$ is an isomorphism. This is the case, for instance, with the ring of formal power series in n variables over a field.

The following change of rings result will be used repeatedly ([<u>20</u>]).

(1.11) <u>Proposition</u>. Let A be a ring, E an A-module, and x an element of A that is a nonzero divisor with respect to both A and E. We have then an isomorphism of functors for $i \geq 0$:

$$\mathrm{Ext}_A^{i+1}(-,E) \cong \mathrm{Ext}^i_{A/(x)}(-,E/xE)$$

in the category of A/(x)-modules.

§1.3 <u>Flatness</u>.

Let A be a ring, E an A-module and

$$(M) \qquad \cdots M_{i+1} \longrightarrow M_i \longrightarrow M_{i-1} \cdots$$

be an arbitrary exact sequence of A-modules. If

$$(M \otimes_A E) \qquad \cdots M_{i+1} \otimes_A E \longrightarrow M_i \otimes_A E \longrightarrow M_{i-1} \otimes_A E \cdots$$

is also exact, we shall say that E is A-flat, or simply flat when the ring is well understood. As $(-)\otimes_A E$ already preserves the right exactness of short exact sequences, the flatness of E amounts to: $(-)\otimes_A E$ preserves injections. If also $M \otimes_A E \neq 0$ whenever $M \neq 0$, then E will be called faithfully flat.

If $\underline{a} = \{a_1, \ldots, a_n\}$ is a sequence of elements of A, the module of relations on the a_i's is the kernel of

$$\phi : A^n \longrightarrow A, \quad \phi(x_1, \ldots, x_n) = \Sigma\, x_i a_i \ .$$

$\mathcal{R}_E(\underline{a}) = \ker(\phi \otimes 1_E)$ is then $\{(e_1, \ldots, e_n) \in E$ with $\Sigma\, a_i e_i = 0\}$ and will be called the module of relations on the a_i's with coefficients in E. Notice that $\mathcal{R}_A(\underline{a})E \subset \mathcal{R}_E(\underline{a})$.

(1.12) Proposition. The following are equivalent for an A-module E.

i) E is flat.

ii) For every sequence $\underline{a}$ as above $\mathcal{R}_A(\underline{a})E = \mathcal{R}_E(\underline{a})$.

Proof. i) $\Longrightarrow$ ii): Let $(\underline{a})$ be the ideal generated by the elements in $\underline{a}$. Consider the exact sequence

$$F \xrightarrow{\psi} A^n \xrightarrow{\phi} A \longrightarrow A/(\underline{a}) \longrightarrow 0$$

where F is a free module. Tensoring with E we obtain the exact sequence

$$F \otimes E \xrightarrow{\psi \otimes 1} A^n \otimes E \xrightarrow{\phi \otimes 1} A \otimes E \longrightarrow A/(\underline{a}) \otimes E \longrightarrow 0.$$

Since $\operatorname{im}(\psi \otimes 1) = \mathcal{R}_A(\underline{a})E$, we have (ii). Actually, in the language of derived functors, the argument shows that

$$\operatorname{Tor}_1^A(A/(\underline{a}),E) = \mathcal{R}_E(\underline{a})/\ \mathcal{R}_A(\underline{a})E\ .$$

ii) $\Longrightarrow$ i): Consider the diagram

$$\begin{array}{ccc} & & P \\ & & \downarrow{\scriptstyle\psi} \\ F & \xrightarrow{\ \phi\ } & G \end{array}$$

where ϕ is injective, P is a free module and ψ is surjective. With $L = \ker \psi$ and $K = \psi^{-1}(\phi(F))$, we obtain the diagram

$$\begin{array}{ccccccc} L \otimes E & \longrightarrow & P \otimes E & \longrightarrow & G \otimes E & \longrightarrow & 0 \\ \uparrow & & \uparrow & & \uparrow & & \\ L \otimes E & \longrightarrow & K \otimes E & \longrightarrow & F \otimes E & \longrightarrow & 0 \end{array}$$

with obvious maps. From it follows that the vertical map on the right will be injective if the two maps ending in $P \otimes E$ are injective. We may then assume that G is a free module. It is also clear that G may be assumed of finite rank. An easy induction on rank takes care of the remainder of the proof, the starting case being supplied by the hypothesis.

It is clear that free modules are flat as are direct limits of free modules. A bit surprising is the converse ([23]): flat modules are direct limits of finitely generated free modules. The localization procedure of §1.2 really amounts to $E_S = E \otimes A_S$; consequently A_S is a flat A-module.

<u>A criterion for flatness</u>:

(1.13) <u>Theorem</u>. Let $\underline{h} : A \longrightarrow B$ be a local homomorphism of the local Noetherian ring $(A,\underline{m})$ into the local coherent ring $(B,\underline{n})$ $(\underline{h}(\underline{m}) \subseteq \underline{n})$. A finitely presented B-module E is flat over A iff

$$\mathrm{Tor}_1^A(A/\underline{m},E) = 0 .$$

Proof. We begin by remarking that if M is a finitely generated A-module then $\mathrm{Tor}_1^A(M,E)$ is a finitely generated B-module. Indeed, if

$$(F) \qquad \cdots \longrightarrow F_n \longrightarrow \cdots F_1 \longrightarrow F_0 \longrightarrow M \longrightarrow 0$$

is a projective resolution of M by finitely generated free A-modules, the complex $F \otimes B$ of B-modules is made up of finitely generated coherent modules and thus its homology groups are finitely generated. According to (1.5 and 1.12) it is enough to show that $\mathrm{Tor}_1^A(A/\underline{p},E) = 0$ for each prime ideal $\underline{p}$. If $\text{K-dim}(A/\underline{p}) = 0$ then $\underline{p} = \underline{m}$ and by hypothesis

$$\mathrm{Tor}_1(A/\underline{p},E) = 0 .$$

Suppose that $\text{K-dim}(A/\underline{p}) > 0$ and that the statement holds for all primes of lower dimension. Let $a \in \underline{m}\ \underline{p}$ and consider the sequence

$$0 \longrightarrow A/\underline{p} \xrightarrow{\cdot a} A/\underline{p} \longrightarrow A/(\underline{p},a) \longrightarrow 0 .$$

Tensoring with E yields

$$\mathrm{Tor}_1(A/\underline{p},E) \xrightarrow{\cdot a} \mathrm{Tor}_1(A/\underline{p},E) \longrightarrow \mathrm{Tor}_1(A/(\underline{p},a),E) .$$

But $\text{K-dim}\,(A/(\underline{p},a)) < \text{K-dim}(A/\underline{p})$ and thus by (1.5) the first module admits a filtration with factors $A/\underline{p}'$, with $\underline{p}'$ properly containing $\underline{p}$; by induction $\mathrm{Tor}_1(A/(\underline{p},a),E) = 0$. But then $\mathrm{Tor}_1(A/\underline{p},E) = a.\mathrm{Tor}_1(A/\underline{p},E)$, which by Nakayama's lemma

(1.6) and the initial remarks forces $\mathrm{Tor}_1(A/\underline{p},E) = 0$.

Remark. Without relative finiteness conditions the vanishing of $\mathrm{Tor}_1(A/\underline{m},E) = 0$ does not ensure the flatness of E as easy examples show.

Let $(A,\underline{m})$ be a local Noetherian ring and E a finitely generated A-module. A topology is induced on E by declaring the submodules $\underline{m}^r\cdot E$ to be a system of neighborhoods for 0. This will be referred to as the $\underline{m}$-adic topology of E.

(1.14) Theorem. (Artin-Rees' lemma) Let F be a submodule of E. There exists an integer $s > 0$ such that for $r \geq s$, we have

$$F \cap \underline{m}^r\cdot E = \underline{m}^{r-s}\cdot(F \cap \underline{m}^s\cdot E).$$

Proof. See [34] for a slightly more general statement.

As consequences we conclude that the $\underline{m}$-adic topology of F is induced from that of E, and, by putting $F = \bigcap_r \underline{m}^r\cdot E$ derive $F = \underline{m}\cdot F$ which by (1.6) implies $F = 0$ --i.e. the $\underline{m}$-adic topology of E is Hausdorff.

Denote by $\hat{E}$ the completion of E with respect to the $\underline{m}$-adic topology; $\hat{E}$ is an $\hat{A}$-module. Actually the canonical map $\hat{A} \otimes E \longrightarrow \hat{E}$ is an isomorphism and $A \longrightarrow \hat{A}$ is a faithfully flat homomorphism ([34]).

The main fact about $\hat{A}$ that we shall use is ([6]):

(1.15) Theorem. (Cohen's theorem) $\hat{A}$ is a homomorphic image of a power series ring $R = D[[x_1,\ldots,x_n]]$ where D is either a field or a complete discrete valuation ring.

§1.4 Fitting's invariants.

Given a finitely generated A-module E we attach to it a sequence of ideals which generalize the classical elementary divisors. Let

$$A^{(\alpha)} \xrightarrow{\phi} A^n \longrightarrow E \longrightarrow 0$$

be a presentation of E with $A^{(\alpha)}$ a free A-module of undetermined rank.

Definition. For an integer $0 \leq r < n$ we call the ideal $F_r(E)$ generated by the minors of order $n-r$ of the matrix ϕ the r-th Fitting ideal of E. We put $F_r(E) = A$ if $r \geq n$.

That $F_r(E)$ does not depend on the presentation follows rather easily from (1.1). Another manner of defining $F_r(E)$ is the following: If $\phi : F \longrightarrow G$ is a homomorphism of A-modules, define the 'order' of ϕ to be the ideal $\underline{o}(\phi) = \Sigma\, f(\phi(F))$ where f runs over $\mathrm{Hom}_A(G,A)$. We could then put in the notion above $\underline{o}(\overset{n-r}{\wedge}\phi) = F_r(E)$ where $\overset{n-r}{\wedge}\phi$ denotes the n-r exterior power of ϕ.

Remark. If $\underline{h} : A \longrightarrow B$ is a ring homomorphism and E is a finitely generated A-module it follows immediately from the first definition that $F_r(E \otimes_A B) = \underline{h}(F_r(E))\cdot B$. (Here the first invariant is viewed as an ideal of B obviously.)

The property that E be a projective module is easily expressed using Fitting's invariants. Thus if E is a module of finite presentation since its projectivity is decided at each localization $E_{\underline{p}}$ we may state ([6]):

(1.16) Proposition. E is projective iff its Fitting's invariants have the property that at each localization $A_{\underline{p}}$ they are either 0 or $A_{\underline{p}}$.

Another property of the $F_r(E)$'s is: If $I_1(E)$ is defined to be the annihilator of E then

$$I_1(E)^n \subseteq F_0(E) \subseteq I_1(E) .$$

More generally we could define the invariant factors of $E : I_r(E)$ being the annihilator of $\overset{r}{\wedge} E$ and derive similar relations. Finally we define still another set of ideals attached to E -- the no name invariants of E. If ξ_r denotes the set of submodules of E that can be generated by r elements, then $K_r(E) = \Sigma$ annihilator E/F, where F runs over the elements of ξ_r.

The relationship between the prime ideals containing these F's, I's and K's is not difficult to determine. Let $\underline{p}$ be a prime ideal and let $E_{\underline{p}}$ be the localization of E at $\underline{p}$. Denote by $\nu(\underline{p};E)$ the minimal number of generators of the $A_{\underline{p}}$-module $E_{\underline{p}}$. It follows easily that $\nu(\underline{p}; \overset{r}{\wedge} E) = \binom{\nu(P;E)}{r}$. We have then

i) $\underline{p} \supseteq I_r$ iff $\nu(P;E) \geq r$.

ii) If $\underline{p} \not\supseteq K_r$, there is $x \in K_r \setminus \underline{p}$ such that $x(E/N) = 0$, where N is a submodule generated by r elements. Localizing at $\underline{p}$ we conclude $E_{\underline{p}} = N_{\underline{p}}$ and $\nu(\underline{p};E) \leq r$. Conversely, if $\nu(\underline{p};E) \leq r$ there is a submodule N generated by r elements and $y \notin \underline{p}$, $yE \subset N$. Thus

$\underline{p} \supseteq K_r$ iff $\nu(\underline{p};E) > r$ and $\mathrm{rad}(K_r) = \mathrm{rad}(I_{r+1})$.

iii) Finally, let $\underline{p} \not\supseteq F_r$; by localizing at $\underline{p}$ and using a previous remark we may take a minimal resolution of E (i.e. the entries of ϕ lie in the maximal ideal of the local ring). The hypothesis then implies $\nu(\underline{p};E) \leq r$. The converse is also clear. Thus $\underline{p} \supseteq F_r$ iff $\nu(\underline{p};E) \geq r+1$.

To sum up : $\mathrm{rad}(F_r) = \mathrm{rad}(K_r) = \mathrm{rad}(I_{r+1})$.

Chapter 2

Homology of Local Rings

The point of view taken in this chapter is that the theory of Macaulay - or Cohen-Macaulay - rings is almost synonymous with studying the homology of local Noetherian rings. Unfortunately little space is devoted to giving ways in which such rings arise in a systematic manner. For some of these aspects we refer to [18, 30] and the bibliographies there.

§2.1 Koszul complexes.

Throughout A will be a commutative ring. We begin with a discussion of what is perhaps the most interesting complex in commutative algebra.

Let E be an A-module and $\wedge(E)$ the exterior algebra of E. For an element $\phi \in \mathrm{Hom}_A(E,A)$ one defines a differential d_ϕ on $\wedge(E)$, in degree n, by the formula

$$d_\phi(e_1\wedge \dots \wedge e_n) = \Sigma(-1)^i\phi(e_i)e_1\wedge \dots \wedge\hat{e}_i\wedge \dots \wedge e_n \quad ;$$

d_ϕ sends $\overset{n}{\wedge}E$ into $\overset{n-1}{\wedge}E$ and, as easily checked, $(d_\phi)^2 = 0$.

When E is a finitely generated module we take, however, an alternate approach.

First we recall the notion of tensor product of chain complexes of A-modules. Let

$$(X,d) : \qquad \dots X_{n+1} \xrightarrow{d_{n+1}} X_n \dots$$

$$(X',d'): \qquad \dots X'_{n+1} \xrightarrow{d'_{n+1}} X'_n \dots$$

be two chain complexes $(X_i = 0,\ i < 0$ always). $(X \otimes X', \partial)$ is defined as

$$(X \otimes X')_n = \bigoplus_i (X_i \otimes X'_{n-i})$$
$$\partial_n = \sum_i (d_i \otimes I + (-1)^{n-i} I \otimes d'_{n-i}) .$$

Let a be an element of the ring A and let A_a be the complex defined as

$$(A_a)_i = 0 \quad \text{for} \quad i \neq 0,1$$
$$(A_a)_i = A \quad \text{for} \quad i = 0,1$$
$$d_1 = \text{multiplication by } a.$$

The Koszul complexes we shall be interested in are built up of such pieces and of modules viewed as complexes in the usual way. Thus if E is an A-module we write

$$E_a = (A_a) \otimes E ,$$

which has as meaningful homology groups:

$$H_0(E_a) = E/aE \quad \text{and} \quad H_1(E_a) = (0 \underset{E}{:} a) = \text{annihilator}$$

of a in E.

Suppose now $\underline{x} = \{x_1,\dots,x_n\}$ is a sequence of elements in A. The Koszul complex associated to $\underline{x}$ is defined as

$$K_\bullet(\underline{x};A) = A_{x_1} \otimes \dots \otimes A_{x_n} .$$

$K_\bullet(\underline{x};A)$ is then the exterior algebra complex associated to $E = A^n$ and the map $\phi : A^n \longrightarrow A$ defined as $\phi(r_1,\dots,r_n) = \sum r_i x_i$.

Finally we shall write $K_\bullet(\underline{x};E)$ for $K_\bullet(\underline{x};A) \otimes E$. The expressions for two of the homology groups of $K_\bullet(\underline{x};E)$ are easily written: $H_0 = E/(x_1,\ldots,x_n)E$ and $H_n = (0 \underset{E}{:} \underline{x})$. In particular if the ideal $(\underline{x})$ does not consist entirely of zero divisors of E, $H_n = 0$, an observation we shall use often. Before we list some technical facts on such complexes notice that $K_\bullet(\underline{x}; -)$ is an exact functor on A-modules.

(2.1) Proposition. Let $C.$ be a chain complex and let $F_\bullet$ be a chain complex of free modules with $F_i = 0$ for $i > 1$. There is then an exact sequence

$$0 \longrightarrow H_0(H_q(C_\bullet)\otimes F_\bullet) \longrightarrow H_q(C_\bullet\otimes F_\bullet) \longrightarrow H_1(H_{q-1}(C_\bullet)\otimes F_\bullet) \longrightarrow 0 .$$

Proof. Construct the sequence of complexes

$$0 \longrightarrow (\hat{F}_0)_\bullet \xrightarrow{f} F_\bullet \xrightarrow{g} (F_1)_\bullet \longrightarrow 0$$

with $(\hat{F}_0)_0 = F_0$, $(\hat{F}_0)_1 = 0$, $(\hat{F}_1)_0 = 0$, $(F_1)_1 = F_1$, f and g the corresponding injection and projection mappings. Tensoring with $C_\bullet$ and writing the homology sequence we get

$$H_{q+1}(C_\bullet\otimes(\hat{F}_1)_\bullet) \xrightarrow{\partial} H_q(C_\bullet\otimes(\hat{F}_0)_\bullet) \longrightarrow H_q(C_\bullet\otimes F_\bullet) \longrightarrow$$
$$H_q(C_\bullet\otimes(\hat{F}_1)_\bullet) \longrightarrow H_{q-1}(C_\bullet\otimes(\hat{F}_0)_\bullet) .$$

Note that $H_{q+1}(C_\bullet\otimes(\hat{F}_1)_\bullet) = H_q(C_\bullet)\otimes F_1$, $H_q(C_\bullet\otimes(F_0)_\bullet) = H_q(C_\bullet) \otimes \hat{F}_0$ and that ∂ is, up to a sign, just the differential of $F_\bullet$. Taking all this into account we obtain the desired sequence.

Write for any complex $C_\bullet$, $C_x = C_\bullet \otimes \Lambda_x$.

(2.2) Proposition. For any chain complex $C_.$, $x\, H(C_x) = 0$.

Proof. Consider A_x and $A_{x,x} = A_x \otimes A_x$. Define chain maps

$$A_x \xrightarrow{f} A_{x,x} \xrightarrow{g} A_x$$

where $f(a) = (a,0)$ and $g(a,b) = a+b$ in dimension 1. This implies a monomorphism

$$H(A_x \otimes C_.) \xrightarrow{f_*} H(A_{x,x} \otimes C_.) .$$

Tensor now

$$0 \longrightarrow (\hat{A}_x)_0 \longrightarrow A_x \longrightarrow (\hat{A}_x)_1 \longrightarrow 0$$

by C_x and take the homology to get the following sequence (2.1):

$$H_q(C_x) \xrightarrow{f_*} H_q(C_{x,x}) \longrightarrow H_{q-1}(C_x) \xrightarrow{(-1)^{q-1}x} H_{q-1}(C_x) .$$

As f_* is a monomorphism, multiplication by x is the null map.

§2.2. Depth.

In this section we shall define a numerical invariant for ideals which plays a role comparable to that of its height. Specifically the situation is as follows: Let I be an ideal and E an A-module. If $x_1 \in I$ is not a zero divisor with respect to E we can ask whether I consists entirely of zero divisors of E/x_1E. In this manner a sequence

$x_1, x_2, \ldots$ of elements of I arises with the property that x_i is not a zero divisor for $E/(x_1,\ldots,x_{i-1})E$. Given noetherian conditions, the sequence $(x_1) \subseteq \ldots \subseteq (x_1,\ldots,x_i) \subseteq \ldots$ stabilizes and eventually I will consist entirely of zero divisors of $E/(x_1,\ldots,x_i,\ldots)E$. We shall refer to $x_1, x_2, \ldots$ as a regular E-sequence, or simply E-sequence.

A first question here is whether the maximal number $\underline{n}$ one obtains is independent of the chosen sequence. Another question is: What is this notion good for? We shall answer the first in the affirmative and give some instances where it can be used in a rather natural way.

(2.3) Theorem. Let $x_1,\ldots,x_n$ be a sequence of elements in A generating an ideal I. Let E be a finitely generated A-module with $E \neq IE$ and let $K_\bullet(\underline{x};E)$ be the corresponding Koszul complex. Let q be the largest integer for which $H_q(K_\bullet) \neq 0$. Then all maximal regular E-sequences in I have the same length $n-q$. We call this number the I-depth of E.

Proof. If $H_n(K_\bullet) \neq 0$, we have by a previous remark $(0 \underset{E}{:} I) \neq 0$ and so all elements of I are zero divisors for E. We use decreasing induction on q. Since $H_0(K_\bullet) = E/IE$, there is a q satisfying our assumption. If $q \neq n$ pick an element a in I which is not a zero divisor of E. Form the exact sequence

$$0 \longrightarrow E \xrightarrow{.a} E \longrightarrow E/aE \longrightarrow 0 ;$$

by the functoriality of the Koszul complex we get

$$H_{q+1}(K_\bullet(\underline{x};E)) \longrightarrow H_{q+1}(K_\bullet(\underline{x};E/aE)) \longrightarrow H_q(K_\bullet(\underline{x};E)) \xrightarrow{.a} H_q(K_\bullet(\underline{x};E)) .$$

From this sequence we have $H_i(K_\bullet(\underline{x};E/aE)) = 0$ if $i > q+1$. By (2.2) $a \cdot H_q(K_\bullet(\underline{x};E)) = 0$ and thus $H_{q+1}(K_\bullet(\underline{x};E/aE)) = H_q(K_\bullet(\underline{x};E))$. Induction ends it.

In the special case where $E = A$ we call I-depth A = <u>grade</u> of I.

(2.4) <u>Theorem</u>. Let J be an ideal of a Noetherian ring A and let E be a finitely generated A-module with $JE \neq E$. If a is in the Jacobson radical of A then (J,a)-depth $E \leq 1 +$ J-depth E.

<u>Proof</u>. First observe that $E/(J,a)E \neq 0$ for otherwise $a(E/JE) = E/JE$, which would contradict Nakayama's lemma. Let $K_\bullet$ be the Koszul complex of E relative to a system of generators of J and $F_\bullet = K_\bullet \otimes \Lambda_a$, that is, $F_\bullet$ is a Koszul complex relative to a system of generators of (J,a). By (2.1) we have

$$0 \longrightarrow H_0(H_q(K_\bullet)\otimes\Lambda_a) \longrightarrow H_q(F_\bullet) \longrightarrow H_1(H_{q-1}(K_\bullet)\otimes\Lambda_a) \longrightarrow 0 .$$

Let q be the integer which determines the J-depth of E. We then have $H_0(H_q(K_\bullet)\otimes\Lambda_a) = H_q(K_\bullet)/aH_q(K_\bullet) \neq 0$ by Nakayama's lemma. The conclusion now follows from (2.3) as $F_\bullet$ is 'longer' by one than $K_\bullet$.

Assume for the remainder of this section that A is Noetherian and modules are finitely generated.

(2.5) Corollary. Let $I = (x_1, \ldots, x_n)$ be contained in the Jacobson radical of A and let E be an A-module. If I-depth $E = n$, then the x's, in any order, form an E-sequence.

Proof. Let $J = (x_1, \ldots, x_{n-1})$; by the previous theorem J-depth $E = n-1$ and by induction $x_1, \ldots, x_{n-1}$ form a regular E-sequence in any order. Let $K_\bullet$ be the Koszul complex corresponding to first $n-1$ generators. Our assumption includes that $H_1(K_\bullet \otimes A_{x_n}) = 0$ (actually it is equivalent) and so, by (2.1) x_n is not a zero divisor of E/JE. Clearly this suffices.

Remarks.

i) If $K_\bullet$ is the Koszul complex associated to a sequence $\underline{x}$ and the A-module E, for any A-algebra B we have $K_\bullet(\underline{y}; E\otimes_A B) = K_\bullet(\underline{x}; E)\otimes_A B$, where the y's are the images in B of the x's. In particular, if B is A-flat we have $H(K_\bullet(\underline{y}; E\otimes B)) = H(K_\bullet(\underline{x}; E))\otimes B$. If applied to the case of localizations we have

$$I\text{-depth } E = \inf \{I_{\underline{p}}\text{-depth } E_{\underline{p}}\}$$

where $\underline{p}$ runs over the primes such that $(E/IE)_{\underline{p}} \neq 0$. Practically this says that in computing the depth we might as well restrict ourselves to local rings.

ii) From the preceding we see that the grade of an ideal does not decrease under localizations. For a prime ideal $\underline{p}$ we shall refer to $\underline{p}_{\underline{p}}$-depth $A_{\underline{p}}$ as the local grade of $\underline{p}$.

Similarity with indeterminates:

The similarity between regular (i.e. A-) sequences and indeterminates is illustrated by the following considerations. Let A be a local Noetherian ring (more generally a coherent ring) containing a field $\underline{k}$ and let $x_1,\dots,x_n$ be a regular A-sequence contained in the maximal ideal $\underline{m}$ of A. We can make A an algebra over $R = \underline{k}[t_1,\dots,t_n]$, the polynomial ring over $\underline{k}$, by letting $t_i \longrightarrow x_i$. $\underline{t} = t_1,\dots,t_n$ being the premier example of a regular sequence, $K_{\bullet}(\underline{t};R)$ is an R-projective resolution of $\underline{k}$. As $x_1,\dots,x_n$ is A-regular, $K_{\bullet}(\underline{t};R) \otimes A = K_{\bullet}(\underline{x};A)$ has trivial homology at dimensions other than 0. In particular $\mathrm{Tor}_1^R(\underline{k},A) = 0$, which according to (1.12) suffices to make A a flat R-algebra. Stated otherwise, the subring $\underline{k}[x_1,\dots,x_n]$ of A is isomorphic to R and A is flat over it.

A consequence is the following: Let I be an ideal of A generated by 'polynomials' in the x_i's with coefficients in $\underline{k}$; then $\mathrm{pd}_A I \leq n$. Indeed, $I = I_0 A$ where I_0 is an ideal of R. By Hilbert's syzygies theorem there is a resolution

$$0 \longrightarrow F_{n-1} \longrightarrow \cdots \longrightarrow F_0 \longrightarrow I_0 \longrightarrow 0$$

where the F_i's are R-free. Tensoring with A yields a projective resolution of I. We can see that $\mathrm{pd}_R I_0 = \mathrm{pd}_A I$.

If A does not contain a field we can still make A an algebra over $S = \mathbb{Z}[t_1,\dots,t_n]$ and similarly conclude

$$\mathrm{Tor}_i^S(\mathbb{Z},A) = 0 \quad \text{for} \quad i > 0 .$$

If characteristic $A = p^n$, we can still conclude

$Tor_i^S(L,A) = 0$, $i > 0$, whenever L is a $\mathbb{Z}$-torsion free S-module. Indeed A is a $\mathbb{Z}_p[t_1,\ldots,t_n]$ module and $Tor_i^S(\mathbb{Z}/p\mathbb{Z},A) = 0$ for $i > 1$, i.e. flat $\dim_S A \leq 1$. Then

$$0 \longrightarrow L \xrightarrow{\cdot p^n} L \longrightarrow L' \longrightarrow 0$$

yields

$$0 = Tor_2^S(L',A) \longrightarrow Tor_1^S(L,A) \xrightarrow{\cdot p^n} Tor_1^S(L,A)$$

and the desired conclusion.

Again a consequence in this case is that if I is an ideal of A generated by polynomials in the x_i's with integral coefficients, and I_0, the corresponding ideal of S is such that S/I_0 is $\mathbb{Z}$-torsion free, then $pd_A I = pd_S I_0$.

If characteristic $A = 0$ then $\mathbb{Z}_p \subset A$ but $\mathbb{Z}_p \cap (x_1,\ldots,x_n)A$ could be $\neq 0$ and thus influence the $\mathbb{Z}$-torsion freeness of $Tor_1^S(L,A)$. There is a case, however, where we might even drop the hypothesis that A be coherent.

Assume $x_1,\ldots,x_n$ form a regular sequence in any order in a commutative ring A and let I be an ideal generated by monomials in the x_i's. Assume also that A is $\mathbb{Z}$-torsion free.

(2.6) Theorem. Let I_0 be the ideal of S in the corresponding monomials in the t's; then $pd_A I \leq pd_S I_0$. Moreover, if $(x_1,\ldots,x_n)$ is contained in rad(I), then $pd_A I = pd_S I_0 = n-1$.

We shall give a proof that includes the statements on the equalities of dimensions in the preceding discussion.

Proof. We show that S/I_0 admits a filtration with factors that are isomorphic to $\mathbb{Z}$. It will follow then that $\mathrm{Tor}_i^S(S/I_0,A) = 0$ for $i > 0$ and whence the finiteness of $\mathrm{pd}_A I$.

The key to the proof is the following ([21]):

(2.7) Lemma. Let $x_1,\ldots,x_n$ be elements constituting an A-sequence in any order. Let J be an ideal generated by monomials in $x_2,\ldots,x_n$. Then $tx_1 \in J$ implies $t \in J$.

Proof. We may assume that x_2 actually occurs in one of the monomials generating J. Write $J = (x_2K,L)$ where K is generated by monomials in $x_2,\ldots,x_n$ and L just by monomials in $x_3,\ldots,x_n$. We have $tx_1 = ux_2 + v$, $u \in K$, $v \in L$. Pass to the ring $A/(x_1)$, noting the homomorphic images of $x_2,\ldots,x_n$ constitute an $A/(x_1)$-sequence. Writing $*$ for homomorphic image we have $u^*x_2^* \in L^*$, whence $u \in (x_1,L)$, say $u = wx_1 \in (K,L)$. We make induction on the sum of the degrees of the monomials generating J and deduce $w \in (K,L)$. Next we substitute for u in the equation $tx_1 = ux_2 + v$ and find $(t-wx_2)x_1 \in L$. Since $x_1,x_3,\ldots,x_n$ is also an A-sequence we have again by induction on n, $t-wx_2 \in L$, hence $t \in (x_2K,L) = J$.

Now use the t's for the x's in this lemma and assume all t's enter in the composition of I_0. Say $I_0 = (t_1K,L)$, where L is generated by monomials in $t_2,\ldots,t_n$. Let $I_1 = (t_1,L)$. Then $I_1/I_0 \cong S/L$. Induction ends easily the proof of the first part of (2.6).

Let us now prove the equality of dimensions. Let

$$0 \longrightarrow P_r \longrightarrow \cdots \longrightarrow P_0 \longrightarrow I_0 \longrightarrow 0$$

be a S-projective resolution of I_0, with P_i finitely generated. Since $I = I_0 \otimes A$ has a finite projective resolution by finitely generated A-projective modules, we have that $\mathrm{pd}_A I = \mathrm{Sup}\,\{\mathrm{pd}_{A_{\underline{p}}} I_{\underline{p}}\}$ $\underline{p}$ running over the prime ideals of A. We may assume A local then and still write for S the localization of $\mathbb{Z}[t_1,\ldots,t_n]$ at the inverse image of the maximal ideal of A. $\mathrm{pd}_S I_0 \leq n-1$ (as $p, t_1,\ldots,t_n$ is a longer S-sequence for any prime $p \in \mathbb{Z}$). As $\mathrm{pd}_S(S/I_0) \geq$ grade I_0 ([20] or (2.24)) we conclude $\mathrm{pd}_S I_0 = n-1$. Assume $\mathrm{pd}_A I < n-1$ and consider a minimal S-resolution for I_0

$$0 \longrightarrow L \longrightarrow F_{n-3} \longrightarrow \cdots \longrightarrow F_0 \longrightarrow I_0 \longrightarrow 0$$

($n > 2$ as lower cases are easily considered); tensoring with A we get $\mathrm{Tor}_1^S(L,A) = 0$ and $L \otimes A$ is a free A-module. Let

$$0 \longrightarrow F_{n-1} \xrightarrow{\phi} F_{n-2} \longrightarrow L \longrightarrow 0$$

be a minimal S-free resolution for L. We get from the above that the Fitting's invariants of L are all 0 or A ((1.16)). Since all the entries of ϕ lie in the maximal ideal of S however, this makes $F_{n-1} \otimes A = \mathrm{Tor}_1(L,A) = 0$, which is a contradiction.

(2.8) Corollary. If $(x_1,\ldots,x_n)$ is a regular A-sequence in any order then $\mathrm{pd}_A(x_1,\ldots,x_n)^r = n-1$.

Relation with Ext:

If A is a commutative ring and I is an ideal of A, we write V(I), the *variety* of I, for the set of primes of A containing I. If M is a module, Supp(M), the *support* of M, will denote the primes $\underline{p}$ for which $M_{\underline{p}} \neq 0$. If M is finitely generated clearly Supp(M) = V(ann(M)).

The following is basic:

(2.9) *Lemma*. If M is a finitely generated module over the Noetherian ring A, then

$$\mathrm{Ass}(\mathrm{Hom}_A(M,N)) = \mathrm{Supp}(M) \cap \mathrm{Ass}(N).$$

Proof. If $\underline{p} \in \mathrm{Ass}(\mathrm{Hom}_A(M,N))$, then as $\mathrm{Hom}_A(M,N)_{\underline{p}} = \mathrm{Hom}_{A_{\underline{p}}}(M_{\underline{p}},N_{\underline{p}}) \neq 0$, $\underline{p} \in \mathrm{Supp}(M)$. If $\underline{p} = \mathrm{ann}(f(M))$, f(M) is a submodule of N and $\underline{p} \in \mathrm{Ass}(N)$.

Conversely, we may assume $\underline{p} \in \mathrm{Supp}(M) \cap \mathrm{Ass}(N)$ is the unique maximal ideal of A. Thus if $n \in N$ is such that $\mathrm{ann}(n) = \underline{p}$,

$$0 \neq \mathrm{Hom}(M/\underline{p}M, An) \hookrightarrow \mathrm{Hom}(M,An) \hookrightarrow \mathrm{Hom}(M,N)$$

and $\underline{p} \in \mathrm{Ass}(\mathrm{Hom}_A(M,N))$.

(2.10) *Corollary*. Let A be a Noetherian ring and let I be an ideal of A and let M be a finitely generated A-module. The following are equivalent:

i) Hom(N,M) = 0 for all finitely generated A-modules N with $\mathrm{Supp}(N) \subseteq V(I)$.

ii) Hom(N,M) = 0 for some finitely generated A-module N with Supp(N) = V(I).

iii) $\mathrm{Ass}(M) \cap V(I) = \emptyset$.

iv) There exists $t \in I$ such that t is M-regular.

An easy induction yields,

(2.11) <u>Corollary</u>. Let A be a Noetherian ring, let I be an ideal of A, let M be a finitely generated A-module and let n be an integer. The following conditions are equivalent:

i) $\mathrm{Ext}^i(N,M) = 0$ for all finitely generated A-modules N with $\mathrm{Supp}(N) \subseteq V(I)$ and all integers $i < n$.

ii) $\mathrm{Ext}^i(N,M) = 0$ for some finitely generated A-module N with $\mathrm{Supp}(N) = V(I)$ and all integers $i < n$.

iii) There exist elements $t_1,\dots,t_n \in I$ forming an M-regular sequence.

§2.3 Macaulay rings.

Assume A to be a Noetherian ring throughout this section. Let I be an ideal of A and let J be an ideal generated by a maximal A-sequence in I; then $I \subset z(A/J)$. In particular some prime ideal containing I will also be contained in $z(A/J)$ and so will have the same grade as I. In other words, grade I = inf {grade $\underline{p}$; $\underline{p}$ minimal over I} . In general, for a prime $\underline{p}$, $\mathrm{ht}(\underline{p}) \geq$ grade $\underline{p}$, for if $x_1,\dots,x_n$ is a regular sequence in $\underline{p}$, $\underline{p}$ contains properly some prime ideal minimal over the ideal generated by $x_1,\dots,x_{n-1}$; but such prime has height at least $n-1$ by induction.

<u>Definition</u>. A Macaulay (or Cohen-Macaulay) ring A is one for which height = grade for each ideal. Similarly an A-module E is said to be a Macaulay module if I-depth E = height(I/J)

for every ideal $I \supseteq \text{ann}(E) = J$; here I/J is viewed as an ideal of A/J.

By the remarks above and the usual definition of height of an ideal we may restrict consideration to prime ideals.

(2.12) <u>Proposition</u>. If $\text{ht}(\underline{p}) = \text{grade } \underline{p}$ for each maximal ideal, then A is a Macaulay ring.

<u>Proof</u>. If $\underline{p}$ is maximum among the primes with grade $\underline{p} <$ $\text{ht}(\underline{p})$, let $\underline{p} \subset \underline{m}$ = maximal ideal; we may assume also that A is a local ring. Let $a \in \underline{m} \setminus \underline{p}$. Then $\text{grade}(\underline{p},a) \leq 1+\text{grade } \underline{p}$ by (2.4), while $\text{ht}(\underline{p},a) \geq 1+\text{ht}(\underline{p})$, thus contradicting $\text{ht}(\underline{p},a) =$ $\text{grade}(\underline{p},a)$.

(2.13) <u>Corollary</u>. If $\underline{p} \subseteq \underline{q}$ are immediate primes (i.e. no primes in-between) in a Macaulay ring A, then $\text{ht}(\underline{q}) = \text{ht}(\underline{p})+1$. In particular two satured chains of primes between two fixed primes have the same length.

<u>Remarks</u>. i) Let A be a Macaulay ring. Then A_S and $A/(x)$ are also Macaulay rings if S is a multiplicative set and x is a nonzero divisor. The power series ring A[[t]] is Macaulay as t lies in the Jacobson radical. As for A[t] the situation is more interesting: Let $\underline{m}$ be a maximal ideal of B = A[t] and localize at $\underline{m} \cap A = \underline{p}$. Then $\text{ht}(\underline{m}) = 1+\text{ht}(\underline{p})$ and $\underline{m} = (\underline{p},f)$, where f may be taken to be monic. But then it is clear that $\underline{p}$-depth A = $\underline{p}$-depth $B/(f)$, since this last module is A-free.

ii) The premier example of a Macaulay ring is a regular ring A: In this case for each maximal ideal $\underline{m}$, $\underline{m}A_{\underline{m}}$ is actually generated by a regular $A_{\underline{m}}$-sequence.

iii) If A is a finitely generated domain over the field $\underline{k}$, by Noether's normalization theorem ([28]) there is a subring R isomorphic to a polynomial ring over $\underline{k}$ such that A is finitely generated as an R-module. It is well-known (e.g.(2.23)) that A Macaulay amounts to saying that A is R-projective.

Now we consider the notion of <u>type</u> of a local Macaulay ring. Let A be a Noetherian ring and let E be an A-module. We quote from [3] the following facts on a minimal injective resolution of E :

$$0 \longrightarrow E \longrightarrow I^0 \longrightarrow I^1 \ldots$$

Each $I^i = \oplus\, I(A/\underline{p})^{\oplus \mu_i(\underline{p})}$ where $I(A/\underline{p})$ is the injective envelope of $A/\underline{p}$ and $\mu_i(\underline{p}) = \dim_{k(\underline{p})} \mathrm{Ext}^i(A/\underline{p},E)_{\underline{p}}$. We shall write $\mu_i(\underline{p};E)$ when mention of the module becomes necessary. Observe that if E is finitely generated then all $\mu_i(\underline{p})$ are finite. Also, in this case, $\mu_i(\underline{p}) = 0$ for $i < \underline{p}$-depth E according to (2.11).

Assume now $E = A$, A a d-dimensional local Macaulay ring. For a prime $\underline{p}$ the first nonzero $\mu_i(\underline{p})$ occurs at $i = \mathrm{ht}(\underline{p}) = r$. We shall call the integer $\mu_r(\underline{p})$ the <u>$\underline{p}$-type</u> of A or the type of A at $\underline{p}$. $\mu_d(\underline{m})$ will be simply called the type of A.

An interpretation of the type of A is obtained in the following manner: Let $x_1,\ldots,x_d$ be a maximal A-sequence in $\underline{m}$; as $\mathrm{Ext}^d(A/\underline{m},A) = \mathrm{Hom}_{A/(\underline{x})}(A/\underline{m},A/(\underline{x}))$, $\mu_d(\underline{m};A) = \mu_0(\underline{m}/(\underline{x});A/(\underline{x}))$; but this last quantity can be interpreted as the number of irreducible components of the $\underline{m}$-primary ideal $(\underline{x})$.

(2.14) <u>Proposition</u>. Let A be a Noetherian ring and let M be a finitely generated A-module. Let $\underline{p}$ be an element of Ass(M), let x be a nonzero divisor relative to M with $M/xM \neq 0$, and let $\underline{m}$ be a prime minimal over $(\underline{p},x)$. Then $\underline{m}$ is associated to M/xM and $\mu_0(\underline{p};M) \leq \mu_0(\underline{m};M/xM)$.

Proof. We may assume that A is local and $\underline{m}$ is its maximal ideal. Let $0:\underline{p}$ be the submodule of M annihilated by $\underline{p}$ and denote by $\underline{p}^*$ the ideal $(\underline{p},x)/(x)$ of $A/(x) = A^*$. Similarly let $0:\underline{p}^*$ be the submodule of $M/xM = M^*$ annihilated by $\underline{p}$. In the surjection

$$M \longrightarrow M^* \longrightarrow 0$$

the elements of $0:\underline{p}$ map into $0:\underline{p}^*$. If $a \in 0:\underline{p}$ maps into 0, $a = xa'$; since $\underline{p}a = x\underline{p}a' = 0$ and x is a nonzero divisor, $a \in x(0:\underline{p})$. Thus we have an injection

$$(0:\underline{p})/x(0:\underline{p}) \hookrightarrow (0:\underline{p}^*).$$

View now this embedding as a sequence of $A/\underline{p}$-modules. Moreover, since $\underline{m}$ is minimal over $(\underline{p},x)$, $A/\underline{p}$ has $\underline{m}/\underline{p}$ as its only non-trivial prime ideal and the modules are Artinian. We now compute approximately their lengths.

(2.15) Lemma. Let A be a local ring with maximal ideal $\underline{m}$ and let I be an $\underline{m}$-primary ideal. Then for any finitely generated module M, $\ell(0:_M I) \leq \ell(A/I)\cdot\ell(0:_M \underline{m})$.

Proof. We use induction on $\ell(A/I)$; if $\ell(A/I) = 1$, $\underline{m} = I$ and there is nothing to show. In any event, A/I contains a sub-module isomorphic to A/I giving rise to an exact sequence

$$0 \longrightarrow A/\underline{m} \longrightarrow A/I \longrightarrow A/J \longrightarrow 0$$

with$\ell(A/I) = \ell(A/J) + 1$. pplying Hom(-,M) to this sequence we get

$$0 \longrightarrow (0:J) \longrightarrow (0:I) \longrightarrow (0:\underline{m})$$

with the annihilator modules taken in M and $\ell(0:I) \leq \ell(0:J) + \ell(0:\underline{m})$.

We apply this to the case of M/xM playing M and $I = (\underline{p},x)$ to get : $\ell(0:\underline{p}) \leq \ell(A/(\underline{p},x))\ell(0:\underline{m}) = \ell(A/(\underline{p},x))\mu_0(\underline{m};M/xM)$.

(2.16) Lemma. Let A be a one-dimensional local domain and let M be a finitely generated A-module. Assume that x is not a zero divisor relative to M. Then $\ell(M/xM) = \ell(A/(x))\ d(M)$, where d(M) is the dimension of the vector space $M_{(0)}$ over K, the field of quotients of A.

Proof. Given M there is a chain

$$M = M_0 \supseteq M_1 \supseteq \ldots \supseteq M_r = 0$$

of submodules such that M_i/M_{i+1} is either A or $A/\underline{m}$. It follows easily that the number of factors isomorphic to A is precisely d(M). Consider the diagram

$$\begin{array}{ccccccccc} 0 & \longrightarrow & M_1 & \longrightarrow & M_0 & \longrightarrow & A/\underline{p} & \longrightarrow & 0 \\ & & .x\downarrow & & .x\downarrow & & .x\downarrow & & \\ 0 & \longrightarrow & M_1 & \longrightarrow & M_0 & \longrightarrow & A/\underline{p} & \longrightarrow & 0 \end{array}$$

where $\underline{p} = 0,\underline{m}$. Assume first $\underline{p} = \underline{m}$; since x is a nonzero divisor relative to M, we have the sequence

$$0 \longrightarrow A/\underline{m} \longrightarrow M_1/xM_1 \longrightarrow M_0/xM_0 \longrightarrow A/\underline{m} \longrightarrow 0$$

and $\ell(M_1/xM_1) = \ell(M_0/xM_0)$, $d(M_1) = d(M_0)$. The analysis of the other case is similar and an easy induction completes the proof

If we apply this to $0:\underline{p}$ in the notation of (2.14) we can complete the proof of the proposition as $\ell((0:\underline{p})/x(0:\underline{p})) = \ell(A/(\underline{p},x)).d(0:\underline{p}) = \ell(A/(\underline{p},x)).\mu_0(\underline{p};M)$.

(2.17) Corollary. The type of a Macaulay ring does not increases under localization.

§2.4 Projective and injective dimensions.

The emphasis in this section is on homological dimensions of finitely generated modules over a local Noetherian ring A. $\underline{m}$ will denote the maximal ideal of A and $\underline{k}$ its residue field.

We begin with some remarks on connected sequences of A-linear functors from mod(A), the category of finitely generated A-modules, into mod(B), where B is a local Noetherian ring and $\underline{h}$: A ⟶ B is a local homomorphism. The functors we have mostly in mind are Ext's and Tor's.

(2.18) Proposition. Let $\{ T^i, i \geq 0 \}$ be a connected sequence of (co)contravariant functors of mod(A). For each $M \in$ mod(A) write $d(M) = \sup \{i \mid T^i(M) \neq 0\}$. Then $d(\underline{k}) = \sup \{d(M)\}$.

We shall refer to $d(M)$ as the T-dimension of M (and will also write $d(M) = T\text{-}\dim(M)$) and to $d(\underline{k})$ as the cohomological dimension of T.

Proof. We give a proof in the contravariant case. If $d(\underline{k})$ is infinite there is nothing to prove. Assume $d(\underline{k}) = s$ is finite and let $n = s+1$. Our claim is that $T^i = 0$ for $i \geq n$. Clearly by the half-exactness of the T's, $T^i(M) = 0$ for a module M of finite length and $i \geq n$. Assume this true for modules of dimension less than r and let $\underline{p}$ be a prime with $\dim(A/\underline{p}) = r$. Pick $x \in \underline{m} \setminus \underline{p}$ and form the exact sequence

$$0 \longrightarrow A/\underline{p} \xrightarrow{\cdot x} A/\underline{p} \longrightarrow A/(\underline{p},x) \longrightarrow 0 .$$

By the connectedness it yields

$$T^n(A/\underline{p}) \xrightarrow{\cdot x} T^n(A/\underline{p}) \longrightarrow T^{n+1}(A/(\underline{p},x)) = 0.$$

The linearity of the T's and Nakayama's lemma forces $T^n(A/\underline{p})=0$.

(2.19) Proposition. Let $\{T^i, i \geq 0\}$ with some $T^i \neq 0$ be a connected sequence of A-linear functors. If $T\text{-dim}(\underline{k}) < \infty$, then for any $M \in \text{mod}(A)$, $T\text{-dim}(M) + \underline{m}\text{-depth } M = T\text{-dim}(\underline{k})$.

Proof. First observe that if $\underline{m}$ is associated to the module M there is a sequence

$$0 \longrightarrow \underline{k} \longrightarrow M \longrightarrow C \longrightarrow 0$$

and $T^r(\underline{k}) \neq 0$ for $r = T\text{-dim}(\underline{k})$ forces, by (2.18), $T^r(M) \neq 0$. If, on the other hand, x is a nonzero divisor relative to M, by Nakayama's lemma we get $T^s(M) \neq 0$ if $s+1 = T\text{-dim}(M/xM)$. After these remarks it follows that $T\text{-dim}(\underline{k}) - T\text{-dim}(M)$ is precisely the length of a maximal M-sequence in $\underline{m}$.

Let E be a finitely generated module and put $T^i = \text{Ext}^i_A(-,E)$. Thus $T\text{-dim}(\underline{k})$ is finite iff E has finite injective dimension. Since $T\text{-dim}(A) = 0$, we have in this case

(2.20) Corollary. $\text{id}_A E = \underline{m}\text{-depth } A = T\text{-dim}(M) + \underline{m}\text{-depth } M$ for any finitely generated module M.

Now we examine the projective dimension of a finitely generated module M using Tor's. Let $r = \text{pd } M$ and let $x_1,\ldots,x_s$ be a maximal M-sequence in $\underline{m}$. It follows easily that $\text{pd } M/(\underline{x})M = r + s$.

(2.21) Proposition. If $\text{pd } M < \infty$ and $\underline{m} \in \text{Ass}(M)$, then pd M = $\underline{m}$-depth A.

Proof. Let $x_1,\ldots,x_n$ be a maximal A-sequence in $\underline{m}$ and $K_\bullet(\underline{x};A)$ the corresponding Koszul complex. But $\mathrm{Tor}_n(A/(\underline{x}),M) = H_n(K_\bullet(\underline{x};A)\otimes M) = 0 \underset{M}{:} (\underline{x}) \neq 0$ and thus pd $M \geq n$.

Before proving the reverse inequality we recall a basic result of the linear algebra over commutative rings ([20]):

(2.22) Proposition (McCoy's theorem). Let A be a commutative ring and let $\phi : M^m \longrightarrow M^n$ be a homomorphism of A-modules given by an n x m matrix (a_{ij}). ϕ is injective iff the ideal generated by the minors of order m does not annihilate a non-zero element of M.

Back to the proof of (2.21) : Let M be a module with pd M = n and let

$$0 \longrightarrow F_n \xrightarrow{\phi_n} F_{n-1} \cdots F_0 \longrightarrow M \longrightarrow 0$$

be a minimal projective resolution of M. Assume that F_n has rank t; let I be the ideal generated by the minors of ϕ_n of order t. We claim that I contains a regular A-sequence of n elements at least. Let x_1 be an element of I which is not a zero divisor of A (achieved through McCoy's theorem). Tensor

$$0 \longrightarrow F_n \xrightarrow{\phi_n} F_{n-1} \cdots F_1 \longrightarrow \mathrm{Im}(\phi_1) \longrightarrow 0$$

by $A/(x_1)$ to get a new exact sequence which represents a minimal projective resolution of the $A/(x_1)$-module $\mathrm{im}(\phi_1)\otimes A/(x_1)$. Induction now ends the proof.

(2.23) Corollary (Equality of Auslander-Buchsbaum). Let A be a local Noetherian ring and let M be a finitely generated

A-module with pd M < ∞. Then pd M + $\underline{m}$-depth M = $\underline{m}$-depth A.

(2.24) Corollary. Let M be a finitely generated module over the Noetherian ring A and let $\underline{p} \in \mathrm{Ass}(M)$. Then $\mathrm{pd}\, M \geq$ grade $\underline{p}$; moreover, if pd M is finite, grade $\underline{p}$ = local grade $\underline{p}$.

Proof. $\mathrm{pd}_A M \geq \mathrm{pd}_{A_{\underline{p}}} M_{\underline{p}}$ = depth $A_{\underline{p}}$ = local grade $\underline{p}$. Now for the proof that depth $A_{\underline{p}}$ = grade $\underline{p}$: Let $x_1,\ldots,x_r$ be a maximal A-sequence in $\underline{p}$. As $\underline{p}$ consists of zero divisors of $A/(\underline{x})$, there is a prime $\underline{q} \supseteq \underline{p}$, $\underline{q} \in \mathrm{Ass}(A/(\underline{x}))$. Localizing at $\underline{q}$ we conclude $\mathrm{pd}\, M_{\underline{q}} = r \geq \mathrm{pd}\, M_{\underline{p}}$ = depth $A_{\underline{p}}$.

The argument also shows that $\underline{p}$ is actually an element of $\mathrm{Ass}(A/(\underline{x}))$. A difficult question in this area is the following (Auslander): Define the grade of a module to be the grade of its annihilator; is it the case that grade M + Krull dim(M) = Krull dim(A)? See [17,30] for a deep discussion of this problem.

§2.5 Euler characteristics of modules.

Here a closer look at the nature of finite projective resolutions is taken. Let A be a commutative ring and let

$$0 \longrightarrow F_n \quad \ldots \quad F_1 \longrightarrow F_0 \longrightarrow M \longrightarrow 0$$

be a finite free resolution of the module M. Define the Euler characteristic of M, $\chi(M)$, to be the integer

$$\chi(M) = \Sigma(-1)^i \mathrm{rank}(F_i).$$

It is easily seen from Schanuel's lemma that $\chi(M)$ does not depend of the chosen resolution (see [20] for the elementary properties of χ that we shall use). Without further ado we use that $\chi(M)$ stays the same for any flat change of rings.

Definition. If $\phi : E \longrightarrow F$ is a homomorphism of A-modules, rank(ϕ) = sup{ $r \mid \overset{r}{\wedge}\phi \neq 0$ }.

If $E = M^{\oplus m}$ and $F = M^{\oplus n}$ are A-modules, we say that $\phi : E \longrightarrow F$ is a M-matricial homomorphism if ϕ is given as $u \otimes 1_M$ where $u : A^m \longrightarrow A^n$.

In this terminology the statement of McCoy's theorem is : a matricial homomorphism $\phi : M^{\oplus m} \longrightarrow M^{\oplus n}$ is injective iff rank$(\phi) = m$ and the ideal generated by the $m \times m$ minors of u does not annihilate a nonzero element of M. Finally, if $u : A^m \longrightarrow A^n$ define $I(u)$ = ideal generated by all $r \times r$ minors of u, r = rank(u).

(2.25) Proposition. Let

$$0 \longrightarrow F_n \xrightarrow{\phi_n} F_{n-1} \cdots F_0 \longrightarrow M \longrightarrow 0$$

be a finite free resolution of the module M. If all $I(\phi_i) = A$ then M is a projective module.

The proof proceeds by showing that F_n splits off F_{n-1} and induction. Note that the converse also holds.

(2.26) Proposition. Let M be a module admitting a finite free resolution. Then

i) $\chi(M) \geq 0$.

ii) $\chi(M) > 0$ iff M is faithful.

iii) $\chi(M) = 0$ iff ann(M) is faithful.

(2.27) Lemma. Let

$$0 \longrightarrow F \xrightarrow{u} G \xrightarrow{v} H$$

be an exact sequence of free A-modules. Then rank(u) + rank(v) = rank(G).

<u>Proof</u>. Tensor the sequence (i.e. change the ring) by A[t] = polynomial ring in t. Then I(u) being a finitely generated faithful ideal acquires a nonzero divisor in A[t]. Localize at the powers of a nonzero divisor of I(u)A[t]; $F \otimes A[t]_S$ then splits off $G \otimes A[t]_S$ and rank(v) = rank(G) - rank(Im(u)). Since the changes of rings leave the ranks unchanged, we have the desired equality.

Proof of (2.26) : Pass to A[t]; the earlier argument shows that all $I(\phi_i)$ are faithful ideals - the case of $I(\phi_n)$ being the statement of McCoy's theorem. If we now localize at the set of regular elements of A[t], the sequence splits piecewise and i),ii) and iii) follow.

(2.28) <u>Corollary</u>. If A is a local Noetherian ring and M is a finitely generated module of finite projective dimension, then M is either faithful or its annihilator contains a nonzero divisor.

We now consider some of the same questions for modules of finite injective dimension over a local Noetherian ring A.

(2.29) <u>Proposition</u>. Let M be a finitely generated module over A of finite injective dimension. Then M is either faithful or its annihilator is faithful.

In Chapter 5 we shall discuss an Euler characteristic for M and a different proof of (2.29). Now we need a few lemmas.

Let M be a nonzero, finitely generated module of finite injective dimension over the local ring A. We saw id M = depth A which is also the maximum value for the depth of any finitely generated A-module according to (2.20) (This fact has provided Hochster with an opening for his solution of various questions on the homology of Noetherian rings [18]). In particular, if id M = 0, A has also Krull dimension 0 as otherwise a module such as $A/\underline{p}$, $\underline{p}$ = minimal prime, would provide an example with depth > 0.

(2.30) Lemma. If id M = 1, then A is a Macaulay ring.

Proof. From (2.20) it follows that A may be assumed to be complete; if $\underline{p}$ is a minimal prime with K-dim A = K-dim$(A/\underline{p})$ and B is the integral closure of $A/\underline{p}$ (which is finitely generated as an $A/\underline{p}$-module ([28]) we have : $\underline{m}$-depth B $\geq$ 2 if K-dim A $\geq$ 2 by [34]. Thus A is one-dimensional.

(2.31) Lemma (Abhyankar-Hartshorne's lemma). Let I and J be nonzero ideals in a commutative ring A such that $I \cdot J = 0$. Then grade (I+J) is at most one.

Proof. We can assume that $I \cap J = 0$ for otherwise, if $0 \neq x \in I \cap J$, then $x(I+J) = 0$. We can even assume A to be a local ring. Let $x = i+j$, $i \in I$ and $j \in J$, be a nonzero divisor; it is clear that $i \neq 0$, $j \neq 0$. Also, $i \notin A(i+j)$ for an equation $i = r(i+j)$ yields $(1-r)i = rj$, which is a contradiction if r is a unit or not. Finally, $(I+J)i \subseteq A(i+j)$, that is grade(I+J) = 1.

In order to apply this to our question, let I be the annihilator of M, let J be the annihilator of I and let J' be the annihilator of J. By the lemma, grade(J+J') $\leq$ 1, in fact = 1. For otherwise I and J would be contained in the same minimal prime $\underline{p}$ of A : $M_{\underline{p}}$ would then be a nonzero, nonfaithful, injective module over the local Artinian ring $A_{\underline{p}}$! Let $\underline{p}$ then be a grade one prime ideal containing J+J'; it is easily seen that $A_{\underline{p}}$ has depth one. We claim that $J_{\underline{p}} \neq 0$ - thus implying that $I_{\underline{p}}$, which is not trivial, consists of zero divisors. But this is indeed the case as $J'_{\underline{p}} \neq A_{\underline{p}}$ is clearly impossible. This is the required reduction. We can make a fresh start and assume that M has injective dimension one over the local ring A of dimension one. There are two cases to examine.

i) $\underline{m}$ is not associated to M.

Here the only primes associated to M are the minimal primes of A containing I. Let

$$0 \longrightarrow M \longrightarrow I^0 \longrightarrow I^1 \longrightarrow 0$$

be a minimal injective resolution of M. I^0 is a direct sum of copies of $I(A/\underline{p})$ = injective envelope of $A/\underline{p}$, for the various primes of height 0 containing I; I^1 is a direct sum of copies of $I(A/\underline{m})$. Let $\underline{p}_1,\ldots,\underline{p}_r$ be the above minimal primes. We can then pick x in some other minimal prime (one containing J) but not in any of the $\underline{p}_i$'s. Map the sequence into itself via multiplication by x and use the snake lemma to get

$$0 \longrightarrow {}_xM \longrightarrow {}_xI^0 \longrightarrow {}_xI^1 \longrightarrow M/xM \longrightarrow I^0/xI^0 \longrightarrow I^1/xI^1 \longrightarrow 0.$$

Since x is not in any of the $\underline{p}_i$'s, it acts as a unit in $I(A/\underline{p}_i)$ and thus ${}_xI^1 = M/xM$. But ${}_xI^1$ is an injective $A/(x)$-module ([4]) and thus $0 \neq M/xM$ is a finitely generated injective module over $A/(x)$. By an earlier remark $A/(x)$ is then an Artinian ring, which is a contradiction as x was taken in a minimal prime.

ii) $\underline{m}$ is associated to M.

We use the notation of i). A minimal injective resolution of M now looks like

$$0 \longrightarrow M \longrightarrow I^0 \oplus E^t \longrightarrow E^s \longrightarrow 0$$

where $E = I(A/\underline{m})$ and the integers t and s are determined by $t = \dim_{\underline{k}}(\mathrm{Hom}(A/\underline{m},M))$ and $s = \dim_{\underline{k}}(\mathrm{Ext}^1(A/\underline{m},M))$. Since $\underline{m}$ is associated to M, $t > 0$. Let $\underline{p}$ be a prime not represented in I^0, that is a minimal prime of J. Applying $\mathrm{Hom}(A/\underline{p},-)$ we get

$$0 \longrightarrow \mathrm{Hom}(A/\underline{p},M) \longrightarrow \mathrm{Hom}(A/\underline{p},I^0 \oplus E^t) \longrightarrow \mathrm{Hom}(A/\underline{p},E^s) \longrightarrow 0$$

exact as $\mathrm{Ext}^1(A/\underline{p},M) = 0$ for depth $A/\underline{p} = 1$. Another way to write this last sequence is

$$0 \longrightarrow {}_{\underline{p}}M \longrightarrow {}_{\underline{p}}I^0 \oplus {}_{\underline{p}}E^t \longrightarrow {}_{\underline{p}}E^s \longrightarrow 0.$$

But ${}_{\underline{p}}I^0 = 0$ as $\underline{p}$ is not contained in any of the primes of I^0. Thus we get

$$0 \longrightarrow {}_{\underline{p}}M \longrightarrow {}_{\underline{p}}E^t \longrightarrow {}_{\underline{p}}E^s \longrightarrow 0,$$

or, in other words, that ${}_{\underline{p}}M$ is a nonzero module of finite length and injective dimension one over $A/\underline{p}$. From the duality theory of [27] it follows that $t = s$.

Define $g(C)$ for any A-module C of finite length to be

$$g(C) = \text{length } \mathrm{Hom}(C,M) - \text{length } \mathrm{Ext}^1(C,M).$$

It is clear that $g(-)$ is additive with respect to short exact sequences. Since $g(\underline{k}) = t-s = 0$ and any module of finite length is an extension of $\underline{k}$'s, we have $g(C) = 0$ for any such module.

Let M_0 be the largest submodule of finite length of M. Also let x be a nonzero divisor of A such that $xM_0 = 0$. $\underline{m}$ is not associated to $M^* = M/M_0 \neq 0$. Using the snake lemma again we get

$$0 \longrightarrow M_0 \longrightarrow {}_xM \longrightarrow {}_xM^* \longrightarrow M_0 \longrightarrow M/xM \longrightarrow M^*/xM^* \longrightarrow 0.$$

But ${}_xM^* = 0$ and thus length(M_0) + length(M^*/xM^*) = length(M/xM). On the other hand the eact sequence

$$0 \longrightarrow A \xrightarrow{\cdot x} A \longrightarrow A/(x) \longrightarrow 0$$

yields that $\mathrm{Hom}(A/(x),M) = M_0$ and $\mathrm{Ext}^1(A/(x),M) = M/xM$. As $A/(x)$ has finite length, $g(A/(x)) = 0$, which implies length$(M^*/xM^*) = 0$, contradicting Nakayama's lemma.

We now return to the examination of the ideals $I(\phi_i)$ in a finite free resolution. The broadest statement on these ideals is contained in the following ([7,30]):

(2.32) Theorem. Let A be a Noetherian ring and let M be a finitely generated A-module. Let

$$M_\bullet : \qquad \ldots M^{r_k} \xrightarrow{\phi_k} M^{r_{k-1}} \ldots$$

be an M-matricial complex with $r_i = 0$ for large i. Then $M_\bullet$ is exact (i.e. $H_i(M_\bullet) = 0$ for $i > 0$) iff

i) $I(\phi_k)$-depth $M \geq k$;

ii) $\mathrm{rank}(\phi_k) + \mathrm{rank}(\phi_{k-1}) = r_{k-1}$.

Proof. If $M_\bullet$ is exact, picking an M-regular element in the 'last' $I(\phi)$ - it exists by McCoy's theorem - and localizing at the multiplicative set it generates we may reduce the length of the complex : ii) would then follow much as for vector spaces; i) is again a direct consequence of McCoy's theorem.

For the converse, the first thing to note is since the ideals $I(\phi)$ contain M-regular elements it is stable under localizations. Let n be the largest integer with $r_n \neq 0$; the statement of ii) makes $\text{rank}(\phi_n) = r_n$ and thus ϕ_n is injective.

By picking an element that is M-regular in $I(\phi_k)$, the statement ii) implies that the homology of $M_\bullet$ is annihilated by an M-regular element. We may assume A local and also that the homology groups all have finite length. We argue by induction on n :

$$0 \longrightarrow M^{r_n} \longrightarrow M^{r_{n-1}} \ldots M^{r_2} \longrightarrow L \longrightarrow 0,$$

$L = \text{image}(\phi_2)$, is then an exact sequence. It also follows that as depth $M \geq n$, depth $L \geq 2$. Let $K = \text{kernel}(\phi_1)$ and consider the sequence

$$0 \longrightarrow L \longrightarrow K \longrightarrow H \longrightarrow 0.$$

If x is an element in the annihilator of H that is M-regular, we get

$$0 \longrightarrow H \longrightarrow L/xL \longrightarrow K/xK \longrightarrow H \longrightarrow 0,$$

and the maximal ideal of A will be associated to L/xL, thus contradicting depth $L/xL \geq 1$, unless $H = 0$.

(2.33) <u>Corollary</u>. Let

$$0 \longrightarrow F_n \cdots \quad F_1 \longrightarrow F_0 \longrightarrow E \longrightarrow 0$$

be a finite free resolution of the module E. Let M be a

finitely generated A-module with the following property: If I is an ideal of grade $r \leq n$, then I-depth M = r. Then the sequence

$$0 \longrightarrow F_n \otimes M \ \dots \ F_1 \otimes M \longrightarrow F_0 \otimes M \longrightarrow E \otimes M \longrightarrow 0$$

is also exact.

§2.6 Gorenstein rings.

We start by looking at the support of a finitely generated A-module of finite injective dimension (A is a Noetherian local ring).

(2.34) Proposition. i) Supp(M) satisfies the equal chain condition - i.e. all maximal chains of primes between two fixed primes have the same length.

ii) K-dim(M) = depth A - inf{ depth $A_{\underline{p}}$, $\underline{p}$ minimal over the annihilator of M}.

Proof. Let $\underline{p}$ be an element of Supp(M); then r = depth $A_{\underline{p}}$ = id $M_{\underline{p}}$ by (2.20). If $\underline{q}$ is a prime immediate over $\underline{p}$ - with K-dim$(A_{\underline{q}}/\underline{p}A_{\underline{q}}) = 1$ - we claim that depth $A_{\underline{q}} = r+1$. Indeed, if $x \subset \underline{q} \setminus \underline{p}$, we have

$$\mathrm{Ext}^r(A/\underline{p},M) \xrightarrow{.x} \mathrm{Ext}^r(A/\underline{p},M) \longrightarrow \mathrm{Ext}^{r+1}(A/(\underline{p},x),M);$$

since $\mathrm{Ext}^r(A/\underline{p},M)_{\underline{p}} \neq 0$, we also get $\mathrm{Ext}^r(A/\underline{p},M)_{\underline{q}} \neq 0$ and thus $\mathrm{Ext}^{r+1}(A/(\underline{p},x),M)_{\underline{q}} \neq 0$ by Nakayama's lemma. This shows that depth $A_{\underline{q}} \geq r+1$.

For the reverse inequality we begin by showing that grade $\underline{p}$ = depth $A_{\underline{p}}$ for any $\underline{p} \in$ Supp(M). Let $\underline{p}$ be maximum with

with the property that depth $A_{\underline{p}} >$ grade $\underline{p}$ and let $x_1,\ldots,x_s$ be a maximal A-sequence in $\underline{p}$. Then $\underline{p}$ is contained in one of the associated primes of $A/(\underline{x})$, say $\underline{q}$. We then have $\mathrm{Ext}^i(A/(\underline{x}),M)_{\underline{q}} = 0$ for $i > s$. By (2.20) $s = \text{depth } A_{\underline{q}} =$ grade $\underline{p}$ and thus by the preceding $\underline{q} = \underline{p}$.

In particular if M is a faithful module K-dim(M) = dim(A) and A is a Macaulay ring. At this point we insert

Bass's <u>conjecture</u> : If a local ring A admits a nonzero finitely generated module M of finite injective dimension, then A is a Macaulay ring.

This question has been settled ([17,30]) for all local rings of equal characteristic and for special modules.

Definition. A local Noetherian ring A is said to be a Gorenstein ring if $\mathrm{id}_A A < \infty$.

It follows that $A_{\underline{p}}$ is also a Gorenstein ring for each prime $\underline{p}$. Global Gorenstein rings could then be defined in the same manner as above or as being locally Gorenstein; clearly these definitions would agree for rings of finite Krull dimension.

If A is an n-dimensional (local) Gorenstein ring, then

i) $$\mathrm{Ext}^n(A/\underline{m},A) \neq 0$$

is the only nonzero of such Ext modules. To examine this module let $x_1,\ldots,x_n$ be a regular A-sequence; by changing the ring (i.e. using (1.11)) we get
$\mathrm{Ext}^n_A(A/\underline{m},A) = \mathrm{Hom}_{A/(\underline{x})}(A/\underline{m},A/(\underline{x}))$ and $\mathrm{Ext}^i_{A/(\underline{x})}(A/\underline{m},A/(x)) = 0$

for i > 0. Thus $A/(\underline{x})$ is a self-injective ring. Since it is indecomposable it must be the injective envelope of each of its nonzero submodules, in particular of $0 :_{A/(\underline{x})} \underline{m}$. Hence $\mathrm{Hom}_{A/(\underline{x})}(A/\underline{m}, A/(\underline{x})) = A/\underline{m}$.

We replace i) by the conditions

ii) A is a Macaulay ring;

iii) $\mathrm{Ext}^n(A/\underline{m}, A) = A/\underline{m}$.

The meaning of iii) becomes clear in a minimal injective resolution of A : that for each prime ideal $\underline{p}$ of height r, then $I(A/\underline{p})$ appears only once in such resolution - as a summand of I^r. Thus A has constant type = 1. Still another way of interpreting iii) is : the ideal $(\underline{x})$ generated by any system of parameters of a Gorenstein ring is irreducible. We may even prove a converse:

(2.35) Theorem. Let A be a local Noetherian ring such that every system of parameters generates an irreducible ideal. Then A is a Gorenstein ring.

Proof. To make an induction on the dimension d of A we first show that $\underline{m}$ contains some nonzero divisor. Let $x_1, \ldots, x_d$ be a system of parameters and let $I_n = (x_1^n, \ldots, x_d^n)$. Then $I_n \neq I_{n-1}$ for otherwise $I_1^{nd} = I_1^{nd-1}$, that contradicts Nakayama's lemma. Now each set $x_1^n, \ldots, x_d^n$ is also a system of parameters and so the I_n's are all irreducible ideals. But the irreducibility of I_n implies that $I_n : \underline{m}$ is contained in all ideals properly containing I_n. Since $I_n = 0$, we have $0 = \cap(I_n : \underline{m}) = 0 : \underline{m}$, i.e. $\underline{m}$ contains some nonzero divisor, say x.

Pass to $A/(x)$; the same hypothesis for systems of parameters is inherited by $A/(x)$.

Remark. It can be shown that although the number of irreducible components of the ideal generated by a system of parameters in a local Macaulay ring is constant the converse does not always hold. The statement above says that if this constant is one then the ring is indeed Macaulay.

In the same spirit of (2.35) is

(2.36) Theorem. Let A be a local Noetherian ring such that for every ideal I generated by a system of parameters I/I^2 is A/I-free. Then A is a Macaulay ring.

Since ideals generated by A-sequences have this 'independence' property, (2.36) is a characterization of Macaulay rings.

Proof. $\dim A = 1$: Let x be a system of parameters (s.o.p. for short); then x^n for all integers n is also a s.o.p.. The assumption is that $(x^n)/(x^n)^2 = A/(x^n)$, that is $(x^{2n}) : x^n = (x^n)$. We claim that x is not a zero divisor: in fact, if $rx = 0$ then $rx^n = 0$ and so $r \in (x^n)$ for all n. By the intersection theorem $r = 0$.

$\dim A > 1$: Assume the statement true for rings of lower dimension. Let $x_1,\ldots,x_d$ be a s.o.p.. Pass to $A' = A/(x_1)$; it is clear that A' inherits the independence property for ideals generated by s.o.p.'s. By induction then $x_2',\ldots,x_d'$ form an A'-sequence. In terms of A this means that

$$(x_1):x_2 = (x_1),\ldots, (x_1,\ldots,x_{d-1}):x_d = (x_1,\ldots,x_{d-1}).$$

Since $x_1^n,\ldots,x_{d-1}^n,x_d$ is also a s.o.p., the relations above hold if for x_i we substitute x_i^n, for $i < d$. Finally, if $rx_d = 0$ we have $r \in (x_1^n,\ldots,x_{d-1}^n)$ and again by the intersection theorem $r = 0$ and x_d is a nonzero divisor. The same can be said of x_1 and so if we add to the relations above $0:x_1 = 0$, we conclude that the x_i's form an A-sequence.

Canonical modules:

(2.37) Theorem. Let A be a local Gorenstein ring. Then any module of finite injective dimension also has finite projective dimension.

We give a proof for finitely generated modules only ([24]). Actually Gorenstein rings can be characterized by this property as in [19].

Proof. Let E be such a module, map a free module over it

$$\text{(S)} \qquad 0 \longrightarrow L \longrightarrow F \longrightarrow E \longrightarrow 0,$$

and note that L also has finite injective dimension. If the depth of E is less than dim A, by (2.11) we have that $\underline{m}$-depth L = 1 + $\underline{m}$-depth E. We can then assume that E has maximum depth already. The claim is that (S) splits. Taking Hom(E,-) of (S) we get

$$0 \longrightarrow \mathrm{Hom}(E,L) \longrightarrow \mathrm{Hom}(E,F) \longrightarrow \mathrm{Hom}(E,E) \longrightarrow \mathrm{Ext}^1(E,L)$$

exact. Since $\underline{m}$-depth E = dim A, from (2.20) it follows that $\mathrm{Ext}^1(E,L) = 0$.

Thus we conclude that H(A), the category of finitely generated module of finite projective dimension, coincides with

I(A), similarly defined for the injective dimension.

Conjecture : If $H(A) \cap I(A) \neq \{0\}$ for a local ring A, then A is a Gorenstein ring.

Given a local ring A, how to find modules of finite injective dimension? In view of Bass's conjecture it is safer to restrict ourselves to Macaulay rings. Let A be a local Macaulay ring; an easy example is obtained in the following manner. Let $\underline{x}$ be a maximal A-sequence in $\underline{m}$ and let E be the injective envelope of $A/\underline{m}$ as an $A/(\underline{x})$-module: then E is a finitely generated A-module and the usual change of rings - viz. (1.11) - shows that it has finite injective dimension. What is much harder is to find examples of higher Krull dimension. There is a case however where this can be done : Assume that A can be written as B/I, with B a local Gorenstein ring (e.g. when A is a localization of an affine algebra)([36]).

(2.38) Theorem. If dim B = n and dim A = d, then $\Omega = \mathrm{Ext}_B^{n-d}(A,B)$ is an A-module of finite injective dimension.

Proof. We prove in fact that $\mathrm{Ext}_A^d(-,\Omega)$ provides a dualizing functor for the category of A-modules of finite length. This fact, with $A = \Omega$, was implicit in our earlier discussion of Gorenstein rings.

Since A is a Macaulay ring of dimension d, we can find a regular B-sequence $\underline{x}$ of length n-d in I. Using (1.11),

$$\Omega = \mathrm{Hom}_{B/(\underline{x})}(B/I, B/(\underline{x}));$$

we may thus assume that dim B = dim A as $B/(\underline{x})$ is also a Gorenstein ring. Then let $T(-) = \mathrm{Ext}_A^d(-,\Omega)$. If

$$0 \longrightarrow M' \longrightarrow M \longrightarrow M'' \longrightarrow 0$$

is a sequence of finitely generated Λ-modules with support in $\{\underline{m}\}$, we may take a system of parameters $\underline{t} = t_1,\ldots,t_d$ in $\underline{m}$ such that $(\underline{t})M = 0$. Notice that $\underline{t}$ is also Ω-regular. Apply the change of rings of (1.11) to have

$$T(M) = \mathrm{Ext}^d_A(M,\Omega) = \mathrm{Hom}_{A/(\underline{t})}(M,\Omega/(\underline{t})\Omega).$$

As $\Omega/(\underline{t})\Omega = \mathrm{Hom}_{B/(\underline{t}')}(A/(\underline{t}),B/(\underline{t}'))$, we finally get $T(M) = \mathrm{Hom}_{B/(\underline{t}')}(M,B/(\underline{t}'))$, where $\underline{t}'$ is a system of parameters for B lifting $\underline{t}$. Since $B/(\underline{t}')$ is self-injective,we are done.

Ω will be called the canonical module of A. In Chapter 4 we discuss the broader class of spherical modules to which Ω belongs. A more extensive theory of 'canonical' modules for rings which are not necessarily Macaulay is found in [16].

(2.39) Corollary. If A is a Macaulay ring that is a UFD and a homomorphic image of a Gorenstein ring, then A is a Gorenstein ring ([29]).

Proof. Let Ω be the module constructed above - not needed here to assume A local. Since A is a domain it is easily seen that Ω may be identified to an ideal of A. As A is a UFD,Ω = dI where d is the g.c.d. of Ω. Changing Ω by I we may assume that grade $\Omega \geq 2$. But in the exact sequence

$$0 \longrightarrow \Omega \longrightarrow A \longrightarrow A/\Omega \longrightarrow 0$$

depth Ω = depth A (for any localization) forces depth $A/\Omega \geq$ dim A - 1, which is a contradiction. Thus Ω is principal and A is a Gorenstein ring.

Remark. The module Ω of (2.38) is as observed a Macaulay module of the same dimension as A. Another property it enjoys is that $\mathrm{Hom}(\Omega,\Omega) = A$: this follows from $\mathrm{Ext}^i(\Omega,\Omega) = 0$ for $i > 0$ and from $\Omega/(\underline{t})\Omega$ being the injective envelope of $A/\underline{m}$ over $A/(\underline{t})$, whenever $\underline{t}$ is a system of parameters.

This last fact has been used ([13]) to show the converse, that is, if Ω is a Macaulay module satisfying $\Omega/(\underline{t})\Omega = I_{A/(\underline{t})}(A/\underline{m})$ for some s.o.p. $\underline{t}$, then A is a homomorphic image of a Gorenstein ring. Proof: Let $B = A \oplus \Omega$ be the trivial extension of A by Ω. The maximal ideal of B is $\underline{m} \oplus \Omega$. To show that B is Gorenstein, first observe that B is a Macaulay A-module; next, by reducing modulo a maximal A-sequence we may assume that A is an Artinian ring - Ω is then the injective envelope of $A/\underline{m}$. To complete the proof it is enough to determine the socle of B : as Ω is a faithful A-module this is evidently $(0, 0\underset{\Omega}{:}\underline{m}) = B/(\underline{m} \oplus \Omega)$.

__How do__ Gorenstein __rings arise__? :

Practically the only way of constructing Gorenstein rings we have discussed is by taking a regular local ring B dividing out by an ideal $I = (x_1,\ldots,x_r)$ generated by a regular B-sequence. In part this difficulty is germane to the nature of Gorenstein rings as B/I with B regular which is Gorenstein, must be a complete intersection if the ideal I is not large.

Let A be a two-dimensional Macaulay ring of type $\mu(A)$. Denote by $\nu(M)$ the minimal number of generators of a module M.

(2.40) __Proposition__. If I is an ideal of grade 2 and projective dimension 1, $\mu(A/I) = \mu(A)(\nu(I)-1)$.

In particular, if I is irreducible it is generated by two elements and A is also a Gorenstein ring.

Proof. Let $0 \longrightarrow F_1 \xrightarrow{\phi} F_0 \longrightarrow I \longrightarrow 0$

be a minimal free resolution of I. Taking $\operatorname{Hom}(\underline{k},-)$ we get

$$\operatorname{Ext}^1(\underline{k},F_1) \xrightarrow{\phi_1} \operatorname{Ext}^1(\underline{k},F_0) \longrightarrow \operatorname{Ext}^1(\underline{k},I) \longrightarrow \operatorname{Ext}^2(\underline{k},F_1) \xrightarrow{\phi_2} \operatorname{Ext}^2(\underline{k},F_0)$$

exact with ϕ_1, ϕ_2 the trivial maps. Thus $\operatorname{Ext}^1(\underline{k},I) = \operatorname{Ext}^2(\underline{k},F_1)$ and also $\operatorname{Ext}^1(\underline{k},I) = \operatorname{Hom}(\underline{k},A/I) = (I:\underline{m})/I = \underline{k}^{\mu(A/I)}$. Finally we get $\mu(A/I) = \mu(A)\cdot\operatorname{rank}(F_1) = \mu(A)(\nu(I)-1)$.

The statement is clearly still true for Macaulay rings of higher dimension - but still with grade I = 2, pd I = 1.

The following is of a more delicate nature:

(2.41) Theorem. Let B be a regular local ring and let I be an ideal of height three such that A = B/I is a Gorenstein ring. Then I is minimally generated by an odd (not always three) number of generators.

Proof. See [8] or [40].

Appendix : Rings of type one.

We may extend the definition of type of a local Macaulay ring given earlier to an arbitrary Noetherian ring A, by saying that the type of A at a prime $\underline{p}$ is

$$\dim_{k(\underline{p})}(\operatorname{Ext}^s_A(k(\underline{p}),A)), \quad s = \text{height } \underline{p}.$$

It can be shown that this number ≥ 1 always ([11]). On the other hand, we saw that for Gorenstein rings the type is

1 at all primes.

Conjecture : Rings of type one are Gorenstein rings.

Of course the difficulty lies in proving that such rings are Macaulay. This is verified here in its simplest case : dim A = 1. Let $(A,\underline{m})$ be such a ring of type one. If

$$0 \longrightarrow A \longrightarrow I^0 \longrightarrow I^1$$

is a minimal injective resolution of A, the hypothesis means

$$I^0 = I(A/\underline{m})^r \oplus A_{\underline{p}_1} \oplus \dots \oplus A_{\underline{p}_n}, \quad I^1 = I(A/\underline{m})$$

where the $\underline{p}_i$'s are the minimal primes of A. The problem lies in proving that $\underline{m}$ is not associated to A, i.e. that $r = 0$.

Step 1 : $r \leq 1$.

Write $I(A/\underline{m}) = E$; apply Hom(-,E) to this sequence to get

$$\mathrm{Hom}(E,E) \xrightarrow{\phi} \mathrm{Hom}(E,E) \oplus \mathrm{Hom}(Q,E) \longrightarrow \mathrm{Hom}(A,E) \longrightarrow 0$$

where we lumped in Q the various $A_{\underline{p}_i}$'s. As $\mathrm{Hom}(E,E) = \hat{A} = \underline{m}$-adic completion of A ([27]), we write

$$\hat{A} \xrightarrow{\phi} \hat{A}^r \oplus \mathrm{Hom}(Q,E) \longrightarrow E \longrightarrow 0, \phi(a) = a(x_1,\dots,x_r,f),\ x_i \in \hat{A}.$$

Call $\phi(\hat{A}) \cap \hat{A}^r = L$; then $\hat{A}^r/L$ embeds in E. Since $\hat{A}^r/L$ is a Noetherian $\hat{A}$-module and E is an Artinian $\hat{A}$-module ([27]), $\hat{A}^r/L$ is a module of finite length. This clearly implies that $r \leq 1$ and, if $r = 1$ that $x = x_1$ is not contained in any minimal prime of $\hat{A}$ as $L \subseteq (x)$.

Step 2 : $r = 0$.

Let y be an element of A such that $y\hat{A} = x\hat{A}$ (possible as

$x\hat{A}$ is an $\underline{m}\hat{A}$-primary ideal). To the presentation apply $\mathrm{Hom}(A/(y),-)$ to get

$$0 \longrightarrow {}_yA \longrightarrow {}_yE \xrightarrow{\cdot x} {}_yE.$$

Since ${}_yE = {}_xE$, we have ${}_yA = {}_yE$. Now apply to the presentation $\mathrm{Hom}(A/(y^2),-)$ to get

$$0 \longrightarrow {}_{y^2}A \longrightarrow {}_{y^2}E \xrightarrow{\cdot x} {}_{y^2}E$$

and then ${}_yA = {}_{y^2}A$. This implies that $0{:}y \hookrightarrow A/(y)$; as ${}_yA = {}_yE$ is an injective $A/(y)$-module, the inclusion above splits. But as $A/(y)$ is indecomposable $(0{:}y) + (y) = A$ is not possible.

Chapter 3

Divisorial Ideals

In this chapter we study some semi-group structures on subsets of ideals of a commutative ring A. The classical example is that of the divisorial ideals of a completely integrally closed domain so successfully used in the study of the factoriality. For other rings this composition has received scant attention and a purpose here is to remedy somewhat this situation. Higher grade ideals also admit a composition but lack thus far any of the properties of grade one divisorial ideals in the sense prescribed in the Preface.

In the second section the emphasis is on the divisor of Auslander-MacRae-Mumford defined on the category of finitely generated torsion modules of finite projective dimension. An exact sequence relating the Grothendieck group of this category to Inv(A) = invertible ideals of A and $\tilde{K}_0(A)$ is useful in this study and serves as a model for the more general treatment of Chapter 4.

The last section is devoted to computations of the length of the torsion part of a module of dimension one, to conditions ensuring the splitting of sequences of such modules, and to an exposition of a result of Burch characterizing the torsion-free modules of rank one and projective dimension one.

§3.1 Composition in Id(A).

Let A be a commutative ring which will be alternatively Noetherian or coherent and denote by Id(A) the set of finitely generated ideals of A that contain a regular element. We will be concerned with various partial compositions in Id(A), that is with group structures - or semi-group ones failing that - that may exist for subsets of Id(A). A key remark will show that there are quite a few of those.

If I is an ideal containing a regular element, $\mathrm{Hom}_A(I,A)$ may be identified with $I^{-1} = \{x \in K \mid xI \subseteq A\}$, where K stands for the total ring of quotients of A. $(I^{-1})^{-1}$ may, in turn, be identified to an ideal of A and $I \subseteq (I^{-1})^{-1}$.

Definition. An ideal I of A is said to be reflexive or divisorial if $I = (I^{-1})^{-1}$.

The reason for the first terminology is clear while the other will soon become so. Notice that as A is coherent, I^{-1} will be a finitely generated fractional ideal of A. To place I and I^{-1} on an equal footing we may then extend the definition of Id(A) to include all regular finitely generated fractional ideals of A. The definition above may then be extended to all such ideals.

For any ideal I, I^{-1} will be reflexive and $(I^{-1})^{-1}$ is the smallest reflexive ideal containing I.

In order to consider a composition on Div(A), the subset of all divisorial ideals in Id(A), we consider first a class of prime ideals of A playing a role in questions of divisibility.

Define $P(A) = \{\underline{p} \in \mathrm{Spec}(A) \mid \underline{p}$ is minimal over an ideal $(a):b$ for some regular element $a\}$.

(3.1) Lemma. Let I be a finitely generated ideal of A

containing a regular element; if $I \not\subset \underline{p}$ for any $\underline{p} \in P(A)$, $I^{-1} = A$. The converse also holds.

Proof. If $x \in I^{-1}$, say $x = b/a$, with a regular, let $\underline{p}$ be a prime ideal minimal over $(a):b$. Then $xI \subseteq A$ implies $I \subseteq \underline{p}$.

Conversely, let $I \subseteq \underline{p}$ = minimal over $(a):b$ and assume $I^{-1} = A$. There is then $s \notin \underline{p}$ and an integer $n > 0$ such that $sI^n \subseteq (a):b$. Pick n least; then $(sb/a)I^{n-1} \cdot I \subseteq A$ and $(sb/a)\ I^{n-1} \subseteq I^{-1} \subseteq A$, a contradiction.

(3.2) Corollary. Let A be an integral domain of quotient field K and let $f = a_n t^n + \ldots + a_0$ be a polynomial in t. $fA[t]$ is a prime ideal iff $fK[t]$ is prime and $(a_n, \ldots, a_0)$ is not contained in any $\underline{p} \in P(A)$. In particular $a+bt$ is prime in $A[t]$ if a,b form a regular A-sequence.

(3.3) Proposition. Two reflexive ideals in Div(A) are equal iff they agree in each $A_{\underline{p}}$, $\underline{p} \in P(A)$.

Proof. Consider the map associated to a reflexive ideal I

$$\phi : A \longrightarrow \Pi\ (A_{\underline{p}}/I_{\underline{p}})\ ,\ \underline{p} \in P(A).$$

Let $x \in \ker(\phi)$ and let $L = \{y \in A \mid yx \in I\}$. Notice that A being coherent L is a finitely generated ideal. Pass the equation $L = I:x$ over to the polynomial ring $A[t]$. We may now add to x an element bt with b regular in I without changing the equation, that is, we may assume that x is regular (No need to assume at this point that $A[t]$ is coherent!). $xL \subseteq I$ yields $x^{-1}L^{-1} \supseteq I^{-1}$; but as $L^{-1} = A$ by (3.1) we conclude $x \in (I^{-1})^{-1} = I$. By faithfully flat descent the conclusion follows.

If I is an element of Id(A) we write D(I) for $(I^{-1})^{-1}$. By going over to A[t] as above we may assume for the purposes here that every ideal in Id(A) has a generating set consisting of regular elements. Thus if $I^{-1} = (d_1, \dots, d_n)$, d_i = regular in K,

$$D(I) = A:I^{-1} = \cap Ad_i^{-1}$$

that says : D(I) is the intersection of all principal ideals containing I. Strictly speaking we can assume the above only after a faithfully flat change of rings.

For two elements I,J ε Div(A), define IoJ = D(IJ).

(3.4) Lemma. The composition 'o' is associative.

Proof. Let $IJ \subseteq Ad$; then $IJd^{-1} \subseteq A$ or $Jd^{-1} \subseteq I^{-1}$ and $J^{-1}d \supseteq (I^{-1})^{-1}$; also $Ad \supseteq JJ^{-1}d \supseteq J(I^{-1})^{-1}$. This says that D(IJ) = D(ID(J)) for any two ideals. Finally apply this to D(D(IJ)K) to get (D(I)oD(J))oD(K) = D(IJK) = D(I)o(D(J)oD(K)).

We shall still note by Div(A) this semi-group structure on the set of divisorial ideals. Let us first describe its invertible elements - A plays the role of the identity.

IoJ = A implies $((IJ)^{-1})^{-1} = A$ or that $(IJ)^{-1} = A$ and from (3.1) IJ is not contained in any prime $\underline{p}$ ε P(A). $(IJ)_{\underline{p}} = A_{\underline{p}}$ says that $I_{\underline{p}}$ is an invertible ideal of $A_{\underline{p}}$ and $J_{\underline{p}}$ is its inverse; also $J = I^{-1}$. We may summarize this in

(3.5) Proposition. I ε Div(A) is invertible (in Div(A)) iff $I_{\underline{p}}$ is an invertible ideal of $A_{\underline{p}}$ for each $\underline{p}$ ε P(A).

In the next statement, if A is not a domain assume that E is an ideal.

(3.6) Proposition. Let A be a coherent local ring where the maximal ideal $\underline{m} \in P(A)$. Let E be a finitely **presented** torsion-less module (i.e. E is a submodule of a product of copies of A); if $E^* = \mathrm{Hom}_A(E,A)$ is A-free then E is A-free.

Proof. Let $J = \mathrm{trace}(E)$; J is a regular ideal of A. If $J = A$, E will admit a summand isomorphic to A and we proceed by induction on the rank of E. If $J \neq A$, let $f_1,\ldots,f_n$ be a basis of E^*; but then $J^{-1}f_1 \oplus \ldots \oplus J^{-1}f_n = E^*$ and $J^{-1} = A$.

(3.7) Corollary. Let A be a Noetherian ring and let E be a finitely generated torsion-less module. Then E is A-projective iff E^* is A-projective and every regular A-sequence of two elements is also E-regular.

Proof. In the sequence

$$0 \longrightarrow E \longrightarrow E^{**} \longrightarrow C \longrightarrow 0$$

$C_{\underline{p}} = 0$ for each $\underline{p} \in P(A)$. Thus $I = \mathrm{ann}(C)$ has grade ≥ 2. Apply $\mathrm{Hom}(A/I,-)$ to get $\mathrm{Hom}(A/I,C) = \mathrm{Ext}^1(A/I,C) = 0$ by (2.11) that is clearly impossible if $I \neq A$.

(3.8) Theorem. $\mathrm{Div}(A)$ is a group iff $A_{\underline{p}}$ is a valuation domain for each $\underline{p} \in P(A)$.

Proof. Write V for $A_{\underline{p}}$ and $\underline{m}$ for $\underline{p}A_{\underline{p}}$. From above it is clear that if all V's are valuation domains $\mathrm{Div}(A)$ is a group. Let us prove the converse by first showing that V is a domain. Let $x \in V$ and let $J = 0:x$; if a is regular in $\underline{m}$, the ideal (J,a) is by (3.6) invertible, say, $(J,a) = Vc$. J is then $= Lc$ and as c is regular, $J = L$ and by Nakayama's lemma

$J = 0$ - since A is coherent J is finitely generated. $\underline{m}$ is, by the same argument a directed union of principal ideals, that is, $\underline{m}$ is a flat ideal.

(3.9) <u>Lemma</u>. Let $(V,\underline{m})$ be a local coherent domain with $\underline{m}$ a flat ideal. Then V is a valuation domain.

<u>Proof</u>. Let x,y be nonzero elements of $\underline{m}$; with $I = (x,y)$ consider a presentation

$$0 \longrightarrow K \longrightarrow V^2 \longrightarrow I \longrightarrow 0.$$

Since the mapping $I \otimes \underline{m} \quad I\underline{m}$ is injective, we get the exact sequence

$$0 \longrightarrow K/\underline{m}K \longrightarrow (V/\underline{m})^2 \longrightarrow I/\underline{m}I \longrightarrow 0$$

showing that as K is finitely generated I must be principal.

The next question to consider is that when divisorial composition is just ordinary multiplication of ideals. This will not be handled here except in the case where $\mathrm{Id}(A) = \mathrm{Div}(A)$. The adjustment to the more general case is not too difficult to perceive.

For an ideal I of A we have the exact sequence

$$0 \longrightarrow A \longrightarrow I^{-1} \longrightarrow \mathrm{Ext}^1(A/I,A) \longrightarrow 0$$

which says that I is reflexive iff I is the annihilator of the module $\mathrm{Ext}^1(A/I,A)$. If x is a nonzero divisor in I, by the usual change of rings $\mathrm{Ext}^1(A/I,A) = \mathrm{Hom}_{A/(x)}(A/I,A/(x)) =$ annihilator of $I/(x)$ in $A/(x)$. Thus if every ideal in $\mathrm{Id}(A)$ is reflexive, $A/(x)$ has the following double annihilator property: For every finitely generated ideal J of $A/(x)$, $0:(0:J) = J$. In the particular case of a Noetherian ring, this would force

A/(x) to be an Artinian ring and actually a Frobenius ring. We shall consider more generally coherent rings.

Assume, by changing notation, that A is the ring A/(x) above. For an ideal I of A we have the sequence

$$0 \longrightarrow \mathrm{ann}(I) = J \longrightarrow A \longrightarrow \mathrm{Hom}(I,A) \longrightarrow \mathrm{Ext}^1(A/I,A) \longrightarrow 0.$$

If $f \in \mathrm{Hom}(I,A)$, $Jf = 0$ means $f(I) \subseteq 0:J = I$. Let us prove that $\mathrm{Hom}(I,I) = A/J$. This is clear if I is a principal ideal; if $I = (L,a)$ assume that the statement holds for L. If ϕ is an element of Hom(I,I), by the preceding $\phi(L) \subseteq L$ and may thus be realized by λ_r i.e. $\phi(x) = rx$ for $x \in L$. Assume the restriction of ϕ to (a) is realized by λ_s. On $K = L \cap (a)$, $\lambda_r(x) = \lambda_s(x)$ and thus $r-s \in 0:K$. But $0:(0:L + 0:a) = (0:(0:L)) \cap (0:(0:a)) = K$. Thus $r-s \in 0:L + 0:a$, say $r-s = p+q$, $p \in 0:L$, $q \in 0:a$. But then λ_{r-p} agrees with ϕ on both L and (a).

(3.10) Theorem. Let A be a local coherent ring of maximal ideal $\underline{m}$ containing a nonzero divisor. If every ideal of Id(A) is reflexive then $\mathrm{Ext}^2_A(E,A) = 0$ for every finitely presented A-module. In particular every finitely presented A-module is reflexive.

Proof. Let E be a finitely presented module. A presentation

$$0 \longrightarrow L \longrightarrow F \longrightarrow E \longrightarrow 0$$

with F finitely generated and free leads to $\mathrm{Ext}^1(L,A) = \mathrm{Ext}^2(E,A)$. Let x be a nonzero divisor in $\underline{m}$; consider the map induced by multiplication by x on **L** : by Nakayama's lemma it is enough to show that $\mathrm{Ext}^2(L/xL,A) = 0$. L/xL admits a filtration

$$L/xL = L_0 \supseteq L_1 \supseteq \ldots \supseteq L_n = 0$$

with factors of the form A/I, I a finitely generated ideal - as A is coherent. As $x \in I$, $I \in Id(A)$. The previous discussion then implies - via change of rings - that $Ext^2(A/I,A) = 0$.

The statement on the reflexivity follows : if

$$A^m \xrightarrow{\phi} A^n \longrightarrow E \longrightarrow 0$$

is a presentation of E and we write $D(E)$ for $\operatorname{coker}(Hom(\phi,1))$ we get the well-known sequence

$$0 \longrightarrow Ext^1(D(E),A) \longrightarrow E \longrightarrow E^{**} \longrightarrow Ext^2(D(E),A) \longrightarrow 0,$$

and the conclusion follows.

(3.10) can also be stated by saying that A has FP (for finitely presented) injective dimension 1.

The following is due to Gruson in another context:

(3.11) <u>Theorem</u>. Let A be a coherent domain of FP-dimension 1. If B = integral closure of A is finitely generated, then B is a Prüfer domain.

<u>Proof</u>. Let $C = Hom_A(B,A)$; C may be identified to an ideal of B and as $Hom_B(-,C) = Hom_A(-\otimes_B B,A) = Hom_A(-,A)$, $Hom_B(-,C)$ is a dualizing functor on the finitely generated torsion-free modules over B (Note that B is a coherent domain).

Let I be a finitely generated ideal of B that we want to prove invertible. Let $J = Hom_B(I,B)$ and note that $Hom_B(J,J) = B$ since B is integrally closed (Only use of the integrally closed condition in proof).

$Hom_B(I,B) = Hom_B(I\otimes C,C) = Hom_B(IC,C)$. Let $T = IC\, Hom_B(IC,C) \subseteq C$; as $Hom_B(T,C) = Hom_B(IC\otimes Hom_B(IC,C),C) = Hom(Hom(IC,C),Hom(IC,C)) = Hom(Hom(I,B),Hom(I,B)) = B$, we conclude T = C.

Let L = I·Hom(I,B), the trace ideal of I. LC = IC Hom(IC,C) = C implies L = A and I is invertible.

In the language of canonical modules:

(3.12) Corollary. If I is a reflexive ideal of a one-dimension Macaulay ring and Hom(I,I) = A, then I is invertible.

Proof. Left as an exercise.

Remark. Let C be an element in Id(A) such that Hom(C,C) = A. If we consider the subset $\mathrm{Div}_C(A)$ of all ideals satisfying J = Hom(Hom(J,C),C), then very much the same theory could be developed for $\mathrm{Div}_C(A)$ instead. This phenomenon will be present in the next chapter.

§3.2 Divisors.

Let $\mathcal{C}$ be a full subcategory of mod(A); we recall that an Euler characteristic on $\mathcal{C}$ is additive mapping

$$\chi : \mathcal{C} \longrightarrow M$$

M an abelian group. Equivalently, χ is a group homomorphism

$$\chi : K_0(\mathcal{C}) \longrightarrow M,$$

$K_0(\mathcal{C})$ the Grothendieck group of $\mathcal{C}$.

To ensure more latitude, we might also consider additive mappings from $\mathcal{C}$ into semi-groups.

Definition. A divisor on $\mathcal{C}$ is an additive mapping χ on $\mathcal{C}$ with values in a semi-group of ideals of A.

These mappings are very reluctant to show up and in these notes we shall consider a few of the more interesting ones that are known. We begin with a discussion of a very fruitful case.

We call a module E over A to be torsion if it is annihilated by a regular element of A. Let $\mathcal{T}$ be the category of finitely generated torsion modules over A that admit a resolution

$$0 \longrightarrow P_n \longrightarrow P_{n-1} \cdots P_0 \longrightarrow E \longrightarrow 0$$

by finitely generated projective modules. Let $K_0(\mathcal{T})$ denote its Grothendieck group. Among the elements of $\mathcal{T}$ are those of the full subcategory $\mathcal{T}_1$ of modules of projective dimension one. Notice that if $E \in \mathcal{T}$ we may map a free $A/(x)$-module E_0 onto E - x here is a regular element in the annihilator of E - and thus derive a resolution

$$0 \longrightarrow E_n \longrightarrow E_{n-1} \cdots E_0 \longrightarrow E \longrightarrow 0$$

with $E_i \in \mathcal{T}_1$. We quote in full the following [5]:

(3.13) Theorem. Let $\mathcal{A}$ be an abelian category and let $\mathcal{P} \hookrightarrow \mathcal{M}$ be full subcategories of $\mathcal{A}$ satisfying the following :

i) $\mathcal{P}$ and $\mathcal{M}$ are closed under finite direct sums.

ii) If $0 \longrightarrow M' \longrightarrow M \longrightarrow M'' \longrightarrow 0$ is exact in $\mathcal{A}$ and $M, M'' \in \mathcal{M}$ then $M' \in \mathcal{M}$.

iii) If M is an object of $\mathcal{M}$, there is an exact sequence

$$0 \longrightarrow P_d \cdots P_0 \longrightarrow M \longrightarrow 0$$

with all $P_i \in \mathcal{P}$.

Then the inclusion $\mathcal{P} \hookrightarrow \mathcal{M}$ induces an isomorphism :

$$K_0(\mathcal{P}) \cong K_0(\mathcal{M}).$$

From the remarks above we then have $K_0(\mathcal{T}_1) \cong K_0(\mathcal{T})$.

Now a look at the torsion modules of projective dimension one. Let

$$0 \longrightarrow K \xrightarrow{u} A^n \longrightarrow E \longrightarrow 0$$

be a presentation of E. As the annihilator of E contains a regular element K has constant rank and is thus a finitely generated module. Let $\underline{d}(E)$ be the ideal generated by the minors of order n of u, that is, $\underline{d}(E) = F_0(E)$. Because K is locally free of rank n, $\underline{d}(E)$ is locally generated by a regular element, i.e. $\underline{d}(E)$ is an invertible ideal of A.

(3.14) Lemma. $\underline{d} : \mathcal{T}_1 \longrightarrow \mathrm{Inv}(A)$ = invertible ideals of A is additive.

Proof. Let $0 \longrightarrow E' \longrightarrow E \longrightarrow E'' \longrightarrow 0$ be an exact sequence in $\mathcal{T}_1$. To show $\underline{d}(E')\cdot\underline{d}(E'') = \underline{d}(E)$ we may assume that A is a local ring and construct resolutions

$$\begin{array}{ccccccccc}
0 & \longrightarrow & F_1' & \longrightarrow & F_1 & \longrightarrow & F_1'' & \longrightarrow & 0 \\
 & & \downarrow{\scriptstyle u'} & & \downarrow{\scriptstyle u} & & \downarrow{\scriptstyle u''} & & \\
0 & \longrightarrow & F_0' & \longrightarrow & F_0 & \longrightarrow & F_0'' & \longrightarrow & 0 \\
 & & \downarrow & & \downarrow & & \downarrow & & \\
0 & \longrightarrow & E' & \longrightarrow & E & \longrightarrow & E'' & \longrightarrow & 0 \\
 & & \downarrow & & \downarrow & & \downarrow & & \\
 & & 0 & & 0 & & 0 & &
\end{array}$$

with the F's free modules. The matrix u is given by

$$u = \begin{bmatrix} u' & * \\ & \\ 0 & u' \end{bmatrix} .$$

As u' and u'' are square matrices we have det(u) = det(u'). det(u''), as desired.

$\underline{d}$ extends then to an additive mapping : $\mathcal{T} \longrightarrow \mathrm{Inv}(A)$ actually defined in the manner :

If
$$0 \longrightarrow E_n \cdots E_0 \longrightarrow E \longrightarrow 0$$
is a resolution of $E \in \mathcal{T}$ by modules $E_i \in \mathcal{T}_1$ then
$$\underline{d}(E) = \underline{d}(E_0)\cdot d(E_1)^{-1} \cdots \underline{d}(E_n)^{(-1)^n}.$$

An immediate consequence is that if $A \longrightarrow B$ is a ring homomorphism with B torsion-free as an A-module, then we have

$$\begin{array}{ccccc} K_0(\mathcal{T}(A)) & = & K_0(\mathcal{T}_1(A)) & \xrightarrow{\underline{d}_A} & \mathrm{Inv}(A) \\ & & \phi \downarrow & & \phi' \downarrow \\ K_0(\mathcal{T}(B)) & = & K_0(\mathcal{T}_1(B)) & \xrightarrow{\underline{d}_B} & \mathrm{Inv}(B), \end{array}$$

where $\phi[E] = [E\otimes B]$ and $\phi'(I) = IB$.

This applies especially to localizations and polynomial extensions.

(3.15) Lemma. For $\underline{p} \in P(A)$ $\mathcal{T}(A_{\underline{p}}) = \mathcal{T}_1(A_{\underline{p}})$.

Proof. It is enough to show that if
$$0 \longrightarrow F \xrightarrow{u} G \longrightarrow M \longrightarrow 0$$
is an exact sequence in $\mathrm{mod}(A_{\underline{p}})$ with F,G finitely generated free and M torsion-free, then it splits. We may assume that all the entries of u lie in the maximal ideal of $A_{\underline{p}}$. Let x be a regular element such that $\underline{p}$ is minimal over $(x):b$. Tensoring the sequence by $A/(x)$ we get $u\otimes A/(x)$ still injective thus contradicting McCoy's theorem.

It follows that for $E \in \mathcal{T}(A)$, $\underline{d}(E)_{\underline{p}}$ is an integral ideal of $A_{\underline{p}}$ for each $\underline{p} \in P(A)$. This forces $\underline{d}(E)$ to be an integral ideal of A ([32]): Write $\underline{d}(E) = Ix^{-1}$ and let $J = \{r \in A \mid rI \subseteq (x)\}$. If $\underline{p}$ is a minimal prime over J we get $\underline{p} \in P(A)$, a contradiction.

Remark. An alternate approach to defining $\underline{d}$ on $\mathcal{T}$ when A is a coherent ring and avoiding the direct use of (3.13) is the following : Define $\underline{d}(E)$ to be the divisorial ideal of A associated to the 0-th Fitting ideal of E, $F_0(E)$. By going over to the various $A_{\underline{p}}$ it follows that $\underline{d}$ provides an additive mapping from $\mathcal{T}$ into the semi-group Div(A). Using now induction on the projective dimension of the module it will follow that the 'o' composition is just plain multiplication. The $\mathcal{T}_1$-resolutions of modules in $\mathcal{T}$ then show that $\underline{d}(E)$ is invertible. The difference in approach is then by defining $\underline{d}(E) = (F_0(E)^{-1})^{-1}$ one starts off with an ideal which we know to be an invariant of E.

(3.16) Corollary. Let I be an ideal of finite projective dimension. Then $I = \underline{d}(A/I)\cdot J$, where J is an overdense ideal, that is $J^{-1} = A$.

Proof. I is the 0-th Fitting ideal of the module A/I and by the definition of $\underline{d}$, I $\underline{d}(A/I)$. We may then express I as a product $\underline{d}(A/I)\cdot J$. Taking inverses twice

$$(I^{-1})^{-1} = \underline{d}(E)\cdot(J^{-1})^{-1} \quad \text{or} \quad (J^{-1})^{-1} = A.$$

(3.17) Theorem. Let A be a local coherent ring in which every finitely generated ideal has finite projective dimension. Then A is a G.C.D. domain.

Proof. That A is a domain follows from (2.26). Let a,b be two nonzero elements of the maximal ideal $\underline{m}$ of A and write I = (a,b). Applying the decomposition above to I we get

$(a,b) = \delta(\alpha,\beta)$ where $(\alpha,\beta)^{-1} = A.\delta$ is the desired gcd.

This is essentially MacRae's theorem ([26]) that says that

in a Noetherian ring every two-generated ideal has projective dimension 0,1 or ∞. In particular it provides a generalization of the factoriality of regular local rings.

We shall discuss the extent to which $\underline{d}$ is universal, that is, an isomorphism. For technical reasons we shall assume A to be Noetherian although with a bit more of care the proofs could be finessed to include coherent rings. We may then take A to be connected - 0,1 are the only idempotents of A.

Denote by $\tilde{K}_0(A)$ the reduced Grothendieck group of the category of finitely generated projective modules over A. For $[P]-[A^n]$ (n = rank(P)), a generator in $\tilde{K}_0(A)$, define

$$\Delta : K_0(A) \longrightarrow K_0(T) \text{ by } \Delta([P]-[A^n]) = [M]-[A/F_0(M)]$$

where $\phi : P \longrightarrow A^n$ is some embedding and $M = \operatorname{coker}(\phi)$. That Δ is a group homomorphism will soon be shown.

(3.18) <u>Theorem</u>. The following is an exact sequence

$$\tilde{K}_0(A) \xrightarrow{\Delta} K_0(T) \xrightarrow{\underline{d}} \operatorname{Inv}(A) \longrightarrow 0.$$

<u>Proof</u>. We begin by verifying that Δ is a group mapping: Let

$$0 \longrightarrow \phi(P) \longrightarrow A^n \longrightarrow M \longrightarrow 0$$

$$0 \longrightarrow \sigma(P) \longrightarrow A^n \longrightarrow N \longrightarrow 0$$

be two embeddings of P into A^n. Since there is a nonzero divisor x such that $x\phi(P) \subseteq \sigma(P)$, we have only to consider the case $\phi(P) \subseteq \sigma(P)$. The sequences above lead to a sequence

$$0 \longrightarrow \sigma(P)/\phi(P) \longrightarrow M \longrightarrow N \longrightarrow 0.$$

Notice that $\sigma(P)/\phi(P) \cong P/\psi(P)$ for some $\psi : P \hookrightarrow P$.

We shall use the following that is based on an idea

of Krämer ([22]):

(3.19) Lemma. If $0 \longrightarrow F \xrightarrow{\phi} F \longrightarrow M \longrightarrow 0$ is exact with F free, then $[M] = [A/\det\phi.A]$ in $K_0(T)$.

Proof. Let (a_{ij}) be an n x n matrix representing ϕ in the canonical basis of F. The ideal generated by the column $(a_{11},\dots,a_{n1})$ is regular by McCoy's theorem. We recall

(3.20) Proposition. Let (a,J) be a regular ideal of the Noetherian ring A. There is then a nonzero divisor of the form a+j, $j \in J$.

Proof. [20].

Apply this to $a = a_{11}$, $J = (a_{21},\dots,a_{n1})$. We may then change the basis of F to make a nonzero divisor to appear in the upper left corner of the matrix for ϕ. Assume then a_{11} to be a nonzero divisor. Let

$$\alpha = \begin{bmatrix} 1 & 0 \dots 0 \\ 0 & a_{11} \dots 0 \\ \dots & \dots \\ 0 & 0 \dots a_{11} \end{bmatrix}$$

and write $\sigma = \alpha\cdot\phi$. We have the exact sequence

$$0 \longrightarrow \operatorname{coker}(\phi) \longrightarrow \operatorname{coker}(\sigma) \longrightarrow \operatorname{coker}(\alpha) \longrightarrow 0$$

and $\det(\sigma) = \det(\alpha)\cdot\det(\phi)$. But $[\operatorname{coker}(\alpha)] = [A/\det\alpha.A]$ is clear while $[\operatorname{coker}(\sigma)] = [A/\det\sigma.A]$ by induction on the rank since the matrix of σ is equivalent to one having all elements in the first column zero but for the first one.

The completion of the proof that Δ is well-defined and a homomorphism is now clear. Since $\underline{d} \circ \Delta = 0$, let us show the exactness of (3.18) by constructing a map

$$\phi : \mathrm{Inv}(A) \longrightarrow K_0(T)/\mathrm{im}\Delta$$

in the following manner: If I is an invertible ideal of A, pick x a regular element of A with $xI \subseteq A$. Put $\phi(I) = [A/xI] - [A/xA]$ in $K_0(T)/\mathrm{im}\Delta$. That Δ is well-defined will again follow from the proof that it is additive.

Let I,J be integral invertible ideals of A. I/JI is a rank one projective module over A/J. Let $\underline{p}_1,\ldots,\underline{p}_n$ be the non-embedded prime ideals of J. Pick b a nonzero divisor in I such that $(b)_{\underline{p}_i} = I_{\underline{p}_i}$ for all i's. Define

$$A/J \xrightarrow{f} I/JI$$

by letting $f(1^*) = b^*$. f is injective as it is so at each prime of Ass(A/J) by construction. Let L = coker(f); in $K_0(T)$ we have $[A/J] + [L] = [I/JI]$. On the other hand, from

$$0 \longrightarrow I/JI \longrightarrow A/JI \longrightarrow A/I \longrightarrow 0$$

we have $[A/JI] = [I/JI] + [A/I]$, and the additivity of ϕ will follow from showing $[L] \in \mathrm{im}\Delta$.

Notice that $\mathrm{ann}(L) \subseteq J$ and as an A/J-module it vanishes when we localize at the elements of Ass(A/J). Since L is a module of projective dimension finite over A/J this means that its annihilator contains a nonzero divisor and thus, as an A-module, grade L $\geq$ 2, and consequently $\underline{d}(L) = A$.

If x is a nonzero divisor in ann(L), we can find a resolution

$$0 \longrightarrow M \longrightarrow F_n/xF_n \ \ldots\ F_0/xF_0 \longrightarrow L \longrightarrow 0$$

with F_i A-free and pd $M \leq 1$. Since $\underline{d}(L) = A$ we get from the additivity of $\underline{d}$ that $\underline{d}(M)$ is a principal ideal. This makes $[A/\underline{d}(M)] - \text{rank}(F_n)[A/xA] + \ldots = 0$. Using this relation with that derived from the sequence above we get
$[L] = \Sigma(-1)^i([F_i/xF_i]-\text{rank}(F_i)[A/xA]) + (-1)^{n+1}([M]-[A/\underline{d}(M)])$,
and $[L] \in \text{im}\Delta$.

(3.21) <u>Corollary</u>. If A is one-dimensional or a local Noetherian ring, then $\underline{d} : K_0(T) \longrightarrow \text{Inv}(A)$ is an isomorphism.

<u>Proof</u>. The local case comes from $\tilde{K}_0(A) = 0$ and the one-dimensional from the proof above.

<u>Remark</u>. The following considerations were pointed out by H.Bass to show that $\underline{d}$ is not in general an isomorphism. Suppose A is a commutative Noetherian ring of finite global dimension. Then we have a commutative exact sequence

$$\begin{array}{ccccc} K_0(T) & \longrightarrow & \tilde{K}_0(A) & \longrightarrow & 0 \\ \underline{d}\downarrow & & \downarrow\text{det} & & \\ \text{Inv}(A) & \longrightarrow & \text{Pic}(A) & \longrightarrow & 0 \\ \downarrow & & \downarrow & & \\ 0 & & 0 & & \end{array}$$

whence an epimorphism $\ker(\underline{d}) \longrightarrow \ker(\det) \longrightarrow 0$. Now $\ker(\det) \neq 0$ in general. For instance, if $A = R[x,y,z]$, $x^2 + y^2 + z^2 = 1$, we have $\text{Pic}(A) = 0$ and $\tilde{K}_0(A) = Z/2Z$.

Let now A be a coherent domain and C denote the category of finitely presented torsion modules over A and let us search for the additive maps from C into Div(A).

(3.22) <u>Theorem</u>. Let $\underline{d} : C \longrightarrow \text{Div}(A)$ be additive and such that $\underline{d}(A/(x)) = (x)$. Then A is integrally closed and $\underline{d}(M) = (F_0(M)^{-1})^{-1}$ for each $M \in C$.

Proof. If $M \in C$ there is a filtration

$$M = M_0 \supseteq M_1 \supseteq \dots \supseteq M_n = 0$$

with $M_i/M_{i+1} \cong A/I_i$, I_i a finitely generated ideal. We then have $\underline{d}(M) = \underline{d}(A/I_0) o \dots o \underline{d}(A/I_{n-1})$.

(3.23) Lemma. If I is an integral ideal in Div(A), then $\underline{d}(A/I) = I$.

Proof. $I = \cap Ax_i$; there is $x \in I$ such that $xx_i \in A$. Thus $xI = \cap Axx_i$ and as $\underline{d}(A/xI) = x\underline{d}(A/I)$ we may assume that I is an intersection of principal integral ideals.

If $y \in I \subseteq Ax$, the surjection $A/(y) \longrightarrow A/I \longrightarrow 0$ leads to $(y) = \underline{d}(A/(y)) = \underline{d}(A/I) o \underline{d}(K)$ and $y \in \underline{d}(A/I)$. On the other hand, $I \subseteq Ax$ leads to a surjection $A/I \longrightarrow A/(x) \longrightarrow 0$ forcing $\underline{d}(A/I) \subseteq \underline{d}(A/(x)) = (x)$. Thus $\underline{d}(A/I) = I$.

The conclusion of (3.22) now follows from (3.8) : The argument above shows $\underline{d}(A/I)$ to be invertible in Div(A) (under 'o') and thus $A_{\underline{p}}$ is a valuation domain for each $\underline{p} \in P(A)$. Since $\underline{d}(-)$ and $(F_0(-)^{-1})^{-1}$ agree at each $A_{\underline{p}}$ they coincide globally.

§3.3 Modules of dimension one.

These modules are studied here particularly with regard to their torsion submodules.

(a) Length of the torsion submodule:

Let $\ell(M)$ denote the length of a module M over the Noetherian ring A. We begin with a consequence of (3.18) [39]:

(3.24) Corollary. Let A be a one-dimensional Noetherian ring and let E be a finitely generated torsion module of projective dimension one. Then $\ell(E) = \ell(A/\underline{d}(E))$.

Proof. Since the elements of $T(A)$ have finite length, $\ell(-)$ is an additive function; as E is equivalent to $A/\underline{d}(E)$ in $K_0(T)$, $\ell(E) = \ell(A/\underline{d}(E))$.

Unfortunately the length of the torsion submodule of a finitely generated module of projective dimension one cannot always be expressed in terms of Fitting's invariants. There are however a few cases where this is possible ([22,39]).

i) Let E be a module presented as

$$0 \longrightarrow A \xrightarrow{\phi} A^n \xrightarrow{\psi} E \longrightarrow 0$$

with $\phi(1) = v = (a_1,\ldots,a_n)$. The torsion submodule $T(E)$ can be determined easily as $\psi(u) \in T(E)$ iff there is a regular element x with $xu = yv$ and $y/xI \subseteq A$ (I is the ideal generated by the a_i's). It follows that $T(E) \cong I^{-1}/A$. Note that $I = F_{n-1}(E)$.

ii) A complementary case is that a torsion module E with a presentation

$$A^{n+1} \xrightarrow{\phi} A^n \longrightarrow E \longrightarrow 0.$$

Since E is a torsion module, we may assume - changing the basis of A^{n+1} if necessary - that ϕ restricted to the submodule generated by the first n basis vectors of A^{n+1} is injective.

If we call L the image in A^n of this submodule, we have

$$0 \longrightarrow \phi(A^{n+1})/L \longrightarrow A^n/L \longrightarrow E \longrightarrow 0.$$

If

$$\phi = \begin{bmatrix} a_{11} & \cdots & a_{1n} & a_{1,n+1} \\ \cdot & \cdots & \cdot & \cdot \\ a_{n1} & \cdots & a_{nn} & a_{n,n+1} \end{bmatrix}$$

we have $\ell(A^n/L) = \ell(A/dA)$ where d is as in the case above the determinant of the left n x n block of ϕ. $\phi(A^{n+1})/L$ on the other hand is a cyclic module A/J with

$$J = \{ r \in A \mid r(a_{1,n+1},\ldots,a_{n,n+1}) \in L \}.$$

By Cramer's rule then $J = (d):F_0(E)$ and finally $\ell(E) = \ell(A/dA) - \ell(A/(d):F_0(E)) = \ell((d):F_0(E)/(d)) = \ell(F_0(E)^{-1}/A)$.

(3.25) Corollary. Let A be an one-dimensional affine domain over a field of characteristic 0 which is a complete intersection. If $\Omega_k(A)$ denotes its module of k-differentials, then $\ell(T(\Omega_k(A))) = \ell(A/J)$, where J is its Jacobian ideal.

Proof. We know that $\Omega_k(A)$ is a module of projective dimension one over A with a resolution (locally):

$$0 \longrightarrow A^n \xrightarrow{\phi} A^{n+1} \longrightarrow \Omega_k(A) \longrightarrow 0.$$

Note $F_1(\Omega_k(A)) = J$. Applying Hom(-,A) to this sequence

$$\mathrm{Hom}(A^{n+1},A) \xrightarrow{{}^t\phi} \mathrm{Hom}(A^n,A) \longrightarrow \mathrm{Ext}^1(\Omega_k(A),A) \longrightarrow 0,$$

and $\ell(\mathrm{Ext}^1(\Omega_k(A),A)) = \ell(A/J)$. Since A is a complete intersection - whence a Gorenstein ring - $\mathrm{Ext}^1(-,A)$ 'reads' the torsion submodule of a module and the conclusion follows.

(b) Splitting the torsion submodule:

Now we take up the question of when the torsion part of a module of projective dimension one splits. Mainly we follow [25]. Let

$$A^m \xrightarrow{\ \phi\ } A^n \longrightarrow E \longrightarrow 0$$

be a presentation of E and let $I_0 \subseteq I_1 \subseteq \ldots$ be the sequence of Fitting's ideals of E. Let I_r be the first nonzero ideal and assume it contains a nonzero divisor of A; this is equivalent to saying that $E \otimes K$ is a projective K-module, K the total ring of quotients of A according to (1.16).

(3.26) Proposition. If I_r is A-projective then pd E $=1$ and its torsion submodule splits.

Proof. We may assume that A is a local ring and if

$$\phi = \begin{bmatrix} a_{11} & \cdots & a_{1m} \\ \cdot & \cdots & \cdot \\ an1 & \cdots & a_{nm} \end{bmatrix}$$

is the matrix of ϕ in the canonical bases, then I_r is generated by the minor d of the $n-r \times n-r$ submatrix sitting in the upper left corner. Let $e_1, \ldots, e_n$ be the corresponding generators of E; from the relations

$$\sum_{j=1}^{n} a_{ij} e_j = 0 \quad (j=1,\ldots,n-r)$$

we get by multiplying by the cofactors of the h-th column and adding

$$de_h + \sum_{j=n-r+1}^{n} d_j e_j = 0 .$$

Since d_j is divisible by d we get that

$$e_h + \sum_{j=n-r+1}^{n} (d_j/d)e_j$$

is annihilated by d and hence lies in T(E). It follows that the images $e^*_{n-r+1},\dots,e^*_n$ of the last r elements generate E/T(E).

If K is the total ring of quotients of A, as observed, $(E/T(E))\otimes K = E\otimes K$ is a free module of rank exactly r. Thus E/T(E) is A-free and T(E) splits off E.

Remark. Actually the hypothesis that I_r be a finitely generated faithful projective ideal suffices to ensure that E has a finite presentation. We would need to modify - for the case of arbitrary commutative ring - the definition of torsion submodule : $e \in E$ is torsion iff ann(e) is a faithful ideal.

Further decomposition of T(E) could perhaps be obtained by requiring that some $F_i(E)$ be also projective for $i > r$. Unfortunately only in the local case or when A is Noetherian of dimension one this can easily be carried out.

(c) A criterion for splitting:

Let A be a one-dimensional Macaulay ring and

$$\text{(S)} \qquad 0 \longrightarrow E' \longrightarrow E \longrightarrow E'' \longrightarrow 0$$

be an exact sequence of modules of projective dimension ≤ 1. If this sequence splits so does the corresponding sequence (beware here) of torsion submodules.

(3.27) Theorem. If A is a Gorenstein ring and $T(E) \cong T(E') \oplus T(E'')$ then the sequence above splits.

At this time we could assume the splitting of

$$0 \longrightarrow T(E') \longrightarrow T(E) \longrightarrow T(E'')$$

or $T(E) \cong T(E') \oplus T(E'')$ arises in some other way. Of course if such a decomposition holds the sequence of torsion submodules is exact at the right by length considerations. We assume the second case to hold in order to present a pretty result of Eisenbud-Hamsher :

(3.28) Theorem. Let $0 \longrightarrow E \longrightarrow F \longrightarrow G \longrightarrow 0$ be an exact sequence of finitely generated modules over the Noetherian ring A. If $F \cong E \oplus G$, then the sequence splits.

Proof. According to [1,Prop.6.5] it is enough to show that $M\otimes(-)$ maintains the sequence exact for any finitely generated module M. We may assume that A is a local ring. If M has finite length $M\otimes F \cong (M\otimes E) \oplus (M\otimes G)$ says that $\ell(M\otimes F) = \ell(M\otimes E) + \ell(M\otimes G)$ and thus

$$M\otimes E \longrightarrow M\otimes F \longrightarrow M\otimes G \longrightarrow 0$$

must be exact on the left. Let now M be an arbitrary (f.g.) module and consider the sequences

$$0 \longrightarrow L \longrightarrow M\otimes E \longrightarrow C \longrightarrow 0$$

$$0 \longrightarrow C \longrightarrow M\otimes F \longrightarrow M\otimes G \longrightarrow 0.$$

Tensor both sequences by $A/\underline{m}^n$ ($\underline{m}$ = maximal ideal of A) to get

$$\begin{array}{ccccccc} (A/\underline{m}^n)\otimes L & \longrightarrow & (A/\underline{m}^n)\otimes(M\otimes E) & \longrightarrow & C/\underline{m}^n\cdot C & \longrightarrow & 0 \\ & & \| & & \downarrow & & \\ & & (M/\underline{m}^n\cdot M)\otimes E & \hookrightarrow & (M/\underline{m}^n\cdot M)\otimes F & & \end{array}$$

by the finite length case. Thus $L \subseteq \underline{m}^n\cdot(M\otimes E)$ for all n. By the intersection theorem $L = 0$.

Proof of (3.27) : The sequence (S) splits if the sequence

$$\mathrm{Ext}^1(E'',E') \longrightarrow \mathrm{Ext}^1(E,E') \longrightarrow \mathrm{Ext}^1(E',E') \longrightarrow 0$$

is exact at the left. As $\mathrm{Ext}^1(-,A)$ is a dualizing functor on torsion modules, the sequence

$$0 \longrightarrow \mathrm{Ext}^1(E'',A) \longrightarrow \mathrm{Ext}^1(E,A) \longrightarrow \mathrm{Ext}^1(E',A) \longrightarrow 0$$

is 'dual' to that of the torsion submodules and thus splits also. Tensoring with E' and taking into account the natural equivalence $\mathrm{Ext}^1(M,A)\otimes(-) \longrightarrow \mathrm{Ext}^1(M,-)$ for a module M (f.g.) of projective dimension one we have the desired end.

Question : Does the same statement in Theorem (3.27) holds for arbitrary 1-dimensional Macaulay rings?

(d) Torsion free modules of projective dimension one and rank one :

Here we give an exposition of a theorem of Burch describing such modules. Let A be a commutative ring and let

$$0 \longrightarrow P \xrightarrow{\ \phi\ } A^n \longrightarrow E \longrightarrow 0$$

be a resolution of E; assume that 0 is the only element killed by a finitely generated faithful ideal of A and that P is a projective module of rank n-1 (This will be the case of a faithful ideal of projective dimension one). By considering a partition of the unity ([6]) assume that $P = A^{n-1}$ and write

$$\phi = \begin{bmatrix} a_{11} & \cdots & a_{1,n-1} \\ \cdot & \cdots & \cdot \\ a_{n1} & \cdots & a_{n,n-1} \end{bmatrix}$$

for the matrix of ϕ. We want to relate E to the ideal generated by the minors of order n-1 of ϕ. Write e_i for the image in E of the i-th canonical basis vector of A^n. The relations

$$\sum_{i=1}^{n} a_{ij} e_i = 0 \quad (j=1,\dots,n-1)$$

yield by Cramer's rule

$$d_i e_j = d_j e_i$$

where d_i is the minor obtained by deleting the i-th row. Let $D = (d_1,\dots,d_n)$ and tentatively define

$$\psi : D \longrightarrow E$$

by $\psi(\Sigma r_i d_i) = \Sigma r_i e_i$. First note that ψ is well-defined as $\Sigma r_i d_i = 0$ yields $(\Sigma r_i d_i) e_k = d_k (\Sigma r_i e_i)$ and thus $\Sigma r_i e_i$ is annihilated by D, a faithful ideal by McCoy's theorem. Thus ψ is an isomorphism and D has also projective dimension one.

(3.29) Lemma. In case E is an ideal of A ψ can be realized by multiplication by an element of A.

Proof. As $\psi \in \mathrm{Hom}(D,A)$, the claim will be established if we prove that the natural homomorphism $A \longrightarrow \mathrm{Hom}(D,A)$ is an isomorphism. To prove equality change the ring to A[t]. As D has finite presentation - being isomorphic to E - we have $\mathrm{Hom}_A(D,A)\otimes A[t] = \mathrm{Hom}_{A[t]}(D[t],A[t])$. In A[t] however D[t] is an ideal containing a nonzero divisor, say $d_1 t + \dots + d_n t^n$. We can now proceed as in [20] and assume that D contains a nonzero divisor already. Hom(D,A) is then D^{-1}. If $a/b \in D^{-1}$, tensoring the resolution of E by A/bA we get

$$0 \longrightarrow (A/bA)^{n-1} \xrightarrow{\phi\otimes A/bA} (A/bA)^n \longrightarrow E/bE \longrightarrow 0$$

exact. Using McCoy's theorem again, the Fitting invariant of order one of E/bE is still faithful, forcing $a \in bA$.

(3.30) Theorem. If $0 \longrightarrow A^{n-1} \xrightarrow{\phi} A^n \longrightarrow I \longrightarrow 0$

is exact and I is an ideal of A, then I = dD where D is the ideal generated by the n-1 x n-1 minors of ϕ.

Appendix : Higher divisorial ideals.

For an ideal I of grade = 1 the divisorial ideal associated with I can be realized as

$$(I^{-1})^{-1} = \mathrm{ann}(\mathrm{Ext}_A^1(A/I,A)).$$

This suggests the following construction for ideals of higher grade.

Definition. $D(I) = \mathrm{ann}(\mathrm{Ext}_A^r(A/I,A))$ where grade I = r.

(3.31) Lemma. If $\underline{a}$ is a A-sequence of length r-1 in I, then $D(I/(\underline{a})) = D(I)/(\underline{a})$.

Proof. This follows immediately from the fact that

$$\mathrm{Ext}_A^r(A/I,A) = \mathrm{Ext}_{A/(\underline{a})}^{r-n}(A/I,A/(\underline{a}))$$

when $\underline{a}$ is an A-sequence of length n contained in I.

Notice that if $\underline{a}$ is an A-sequence of length r-1 in I, where grade I = r, then $D(I/(\underline{a})) = ((I/(\underline{a}))^{-1})^{-1}$ which is an intersection of principal fractional ideals of $A/(\underline{a})$ -indeed finitely many of them. With this in mind we consider an A-submodule of the quotient ring of A

$$L = \bigcap_i t_i^{-1}(\underline{a}_i)$$

where $\underline{a}_i$ is an A-sequence of length r and $t_i^{-1}(\underline{a}_i) \supseteq I$. We would like to show that L = D(I) and we may assume that I = D(I) since D(I) = D(D(I)).

Suppose $\underline{a}$ is an A-sequence in I of length = r-1. If $(\bar{u}/\bar{t})(A/(\underline{a}))$ is a principal fractional ideal of $A/(\underline{a})$ containing $I/(\underline{a})$, then $t^{-1}(\underline{a},u) \supseteq I$ with t and u suitably chosen inverse images of $\bar{t}$ and $\bar{u}$, with t chosen to be a nonzero divisor by prime avoidance. Moreover $\underline{a},u$ is an A-sequence and either $\underline{a},t$ is an A-sequence or t is a unit. If we let T be the intersection of all such fractional ideals of A, then T contains L. Furthermore, $T \cap A/(\underline{a}) = I/(\underline{a})$ since $I/(\underline{a})$ is reflexive. But $(\underline{a})$ is contained in both $T \cap A$ and I so $T \cap A = I$. Hence we would be through if we could show that L is contained in A.

If any of the components of the intersection for L is contained in A we are through, so we assume this is not the case. Consider $t^{-1}(\underline{a},u)$ as above. Now $\underline{a},t$ is an A-sequence and again by prime avoidance we can find a nonzero divisor v such that $\underline{a},v$ is an A-sequence, $v,\underline{a}$ is an A-sequence, and $v = t \bmod(\underline{a})$. This last condition insures that $v^{-1}(\underline{a},u) \supseteq I$. We can also find a nonzero divisor of the form $s = v+b$ where $b \in (\underline{a})$ and v,s is an A-sequence. Since $t^{-1}(\underline{a},u) \supseteq s^{-1}(\underline{a},u) \supseteq L$, we find that $(v,s) \subseteq L^{-1}$. Taking inverses again we obtain $(v,s)^{-1} \supseteq (L^{-1})^{-1} \supseteq L$. But $(v,s)^{-1} = A$ since grade(v,s) = 2. Therefore we have proved that L is contained in A as desired, that is :

(3.32) <u>Theorem</u>. L = D(I).

Chapter 4

Spherical Modules and Divisors

One of the aims of this chapter is to study the modules that can be obtained as homomorphic images of direct sums of copies of a fixed module G. A result of Gruson says that if G is finitely generated and faithful the category mod(A) comes close to being so obtained, as each module M admits a finite filtration with factors that are images of sums of copies of G. Applications of this are made to change of rings and homological dimensions.

Calling C(G) the category generated by G, we turn to the examination of the modules in C(G) that admit resolutions by direct sums of G's, or more especially finite resolutions. The situation becomes then ripe for applying the procedure of Chapter 3 and assigning a divisor to such modules. Expectedly some restrictions must be placed- to obtain a degree of invariance - on the module G. A class of modules - spherical ones - for which the procedure works optimally is discussed.

§4.1 A theorem of Gruson.

(4.1) Theorem. Let E be a finitely generated faithful module over the commutative ring A. Then every A-module admits a finite filtration of submodules whose factors are quotients of direct sums of copies of E.

It is as if every finitely generated faithful module is

'piecewise' a generator for the category mod(Λ).

Proof. It is enough to show this for Λ itself : Let $I_0 \subseteq I_1 \subseteq \ldots \subseteq \Lambda$ be a sequence of ideals of Λ such that for each i there is a surjection

$$\phi_i : E^{(L_i)} \longrightarrow I_i/I_{i-1} .$$

Let $\phi : A^{(L)} \longrightarrow M \longrightarrow 0$ be a presentation of the module M. Then the series of submodules of M given by $\phi(I_i^{(L)})$ has the desired properties.

For the proof of the ring A itself we use the Fitting invariants of E. Let

$$A^{(L)} \xrightarrow{u} A^n \longrightarrow E \longrightarrow 0$$

be a free presentation of E and denote by (a_{ij}), $1 \leq j \leq n$, $i \in L$, the matrix of u. Let $F_r(E)$ be the ideal generated by the minors of order n-r of this matrix.

The smallest integer r such that $F_r \neq 0$ exists and is nonzero; the smallest integer m such that $F_m = A$ exists and $m \geq r$. We notice that $m = r$ means that E is a projective module ((1.16)) and in this case we are done.

We shall reason by induction on $d = m-r$.

Suppose $d > 0$. Let $I = \text{ann}(E/F_rE)$. We can apply the induction hypothesis to the Λ/I-module E/IE : (a) this module is faithful ($xE \subset IE \subset F_rE$ implies $x \in I$), and (b) its r-th Fitting ideal is trivial and its m-th ideal is Λ/I.

We have also that $I^n \subseteq F_r$. This gives rise to a filtration

$$F_r \subseteq F_{r+1} + I^{n-1} \subseteq \ldots \subseteq I$$

where each factor is an Λ/I-module. Thus to complete the proof it is enough to show that F_r is a quotient of a direct sum of copies of E. Let D be the determinant of the $n-r = s$ minor defined by (a_{i_k,j_ℓ}), $1 \leq k,\ell \leq s$, and let us look for a linear form on E whose image contains D. One chooses an integer t between 1 and n and distinct from the i_k's : this is possible since $r > 0$. We consider the following linear form on A^n

$$(x_1,\ldots,x_n) \longrightarrow \det\begin{bmatrix} a_{i_1,j_1} & \cdots & a_{i_1,j_s} & x_1 \\ \cdot & \cdots & \cdot & \cdot \\ a_{i_s,j_1} & \cdots & a_{i_s,j_s} & x_s \\ a_{i_t,j_1} & \cdots & a_{i_t,j_s} & x_t \end{bmatrix}$$

As the minors of order $s+1$ are zero, this form is trivial on the image of u; it defines then, by passage to the quotient, a linear form on E. On the other hand, it transforms the t-th vector of the canonical basis of A^n into D.

We point out some immediate consequences.

(4.2) Corollary. Let F be a right exact functor which commutes with arbitrary direct sums and is trivial on E: then F is trivial.

(4.3) Corollary. Let E be a finitely generated faithful module over the commutative ring A and let M be an A-module. If $E \otimes M = 0$, then $M = 0$.

§4.2 Change of rings and dimensions.

The change of rings problem in homological dimension theory is the following:

Let $f : A \longrightarrow B$ be a homomorphism of rings and let E be a module over B. How are $pd_A E$ and $pd_B E$ (or some other dimension) related? In general one has that

$$(1) \qquad pd_A E \leq pd_A B + pd_B E.$$

This follows, for instance, from the spectral sequence ([10])

$$(2) \qquad E_2^{p,q} = Ext_B^p(E, Ext_A^q(B,C) \underset{p}{\longrightarrow} Ext_A^n(E,C)$$

where C may be any A-module. If $d = pd_A B$ and $e = pd_B E$ are both finite, there are several instances where this spectral sequence allows one to conclude that $E_2^{d,e} \neq 0$, and hence equality in (1) above.

Here we will study the case where B is not only finitely generated as an A-module but admits a resolution

$$(3) \qquad 0 \longrightarrow P_d \quad \cdots \quad P_1 \longrightarrow P_0 \longrightarrow B \longrightarrow 0$$

where the P_i's are finitely generated projective modules.

In this case the module $Ext_A^d(B,A)$ plays a significant role and we begin with an elementary discussion of it. A first property to notice is that if C is any A-module we have a canonical isomorphism

$$Ext_A^d(B,C) \cong Ext_A^d(B,A) \otimes_A C.$$

This arises either from a spectral sequence argument or more simply from the following considerations. As only the module structure of B is involved we may assume that pd B = 1. Write then a resolution

$$0 \longrightarrow P_1 \longrightarrow P_0 \longrightarrow B \longrightarrow 0$$

with P_0, P_1 finitely generated projective modules. Applying $Hom_A(-,C)$ we get

$$\mathrm{Hom}_A(P_0,C) \longrightarrow \mathrm{Hom}_A(P_1,C) \longrightarrow \mathrm{Ext}^1_A(B,C) \longrightarrow 0.$$

The result now follows from observing the natural equivalence $\mathrm{Hom}_A(P_i,-) = \mathrm{Hom}_A(P_i,A)\otimes_A(-)$, as P_i is projective and finitely generated.

We are going to consider two types of conditions on B: for both the module $T = \mathrm{Ext}^d_A(B,A)$ plays an essential role, and which are, under general circumstances, necessary for equality in (1). This situation we shall, at times, abbreviate by saying that there is change of rings.

(a) Macaulay <u>extensions</u> :

We assume here that

(Condition $\mathcal{M}$) $\mathrm{Ext}^i_A(B,A) = 0$ for $i < d$.

This restriction is reminiscent of the condition a factor ring of a regular local ring must satisfy to be a Macaulay ring.

In this case, if we apply $\mathrm{Hom}_A(-,A)$ to the sequence (3) twice, taking into account the reflexivity of the projective modules, and the maps between them, we obtain that $\mathrm{Ext}^d_A(T,A) \cong B$. In particular, we conclude that T is a faithful module.

(4.4) <u>Theorem</u>. Let B be an extension of A satisfying conditions (3) and ($\mathcal{M}$) above. Then for any B-module E with $\mathrm{pd}_B E = e < \infty$, we have $\mathrm{pd}_A E = \mathrm{pd}_A B + \mathrm{pd}_B E$.

<u>Proof</u>. According to (2) it will be enough to show that for some $C = A^{(L)}$

$$\mathrm{Ext}^e_B(E,\mathrm{Ext}^d_A(B,C)) = \mathrm{Ext}^e_B(T\otimes_A C) = \mathrm{Ext}^e_B(E,T^{(L)}) \neq 0.$$

Remark. For the injective and flat analogues of (4.4) one can proceed in a similar way to show the equality in the dimension formula using the following spectral sequences

$$E_2^{p,q} = \mathrm{Ext}_B^p(\mathrm{Tor}_q^A(B,C),E) \underset{p}{\longrightarrow} \mathrm{Ext}_A^n(C,E)$$

and

$$E_2^{p,q} = \mathrm{Tor}_p^B(E,\mathrm{Tor}_q^A(B,C)) \underset{p}{\longrightarrow} \mathrm{Tor}_n^A(C,E)$$

respectively.

Let us indicate how to proceed in the injective case (one uses the same type of argument - and module - in the flat case. Let M be an injective A-module. Because $\mathrm{Ext}_A^i(B,A) = 0$, $i < d$, it follows much in the same way as before that $\mathrm{Tor}_d^A(B,M) = \mathrm{Hom}_A(T,M)$. Let $\mathrm{Ext}_B^e(C,E) \neq 0$ for some B-module C; then $\mathrm{Ext}_B^e(T,E) \neq 0$ also.

Let $0 \longrightarrow T \longrightarrow M$ be an A-injective envelope of T. $\mathrm{Ext}_A^{d+e}(M,E) = 0$ implies that $\mathrm{Ext}_B^e(\mathrm{Hom}_A(T,M),E) = 0$. Since T is finitely generated and faithful as a B-module, there is an exact sequence

$$0 \longrightarrow B \longrightarrow T^n$$

obtained by mapping $1 \longrightarrow (x_1,\ldots,x_n) \in T^n$, where the x_i's form a generating set for T. But in this case $\mathrm{Ext}_B^e(\mathrm{Hom}_A(B,M),E) = 0$ also and then $\mathrm{Ext}_B^e(\mathrm{Hom}_A(B,T),E) = 0$. As there is a surjection

$$\mathrm{Hom}_A(B,T) \xrightarrow{\phi} T \longrightarrow 0$$

given by $\phi(f) = f(1)$ we contradict $\mathrm{Ext}_B^e(T,E) \neq 0$.

(b) Finitely generated modules :

If E itself, in the previous considerations, is finitely

generated there is not a need to impose such a strong restriction on B.

(4.5) Theorem. Let A and B be rings, with B local and finitely generated as an A-module. Assume that B admits a finite resolution by finitely generated projective A-modules. Let E be a B-module admitting a finite free resolution over B. Then $pd_A E = pd_A B + pd_B E$.

This result, contrary to (4.4) has an elementary via of attach through the device of minimal projective resolutions.

Proof. Using the previous notation notice that

$$Ext_B^e(E, Ext_A^d(B,A)) = Ext_B^e(E,B) \otimes_B Ext_A^d(B,A)$$

as E has a finite presentation. Since both modules in the product are nonzero and the ring B is local, the statement follows.

(c) Some equalities of dimensions:

Before we derive other consequences of (4.4) and (4.5) recall the definition of the finitistic projective dimension of a ring A - FPD(A) for short - : FPD(A) is obtained by taking in the definition of the global dimension of A only those modules of finite dimension. Moreover, if we only consider the finitely generated modules we obtain fPD(A).

(4.6) Theorem. Let A be a coherent local ring and let I be a finitely generated ideal with $pd_A I < \infty$. Then

$$fPD(A) = pd_A(A/I) + fPD(A/I).$$

When A is Noetherian there is a simple way - via the complementary notion of depth - to verify this equality.

For the rest of this section we shall assume all unspecified A-modules to be of finite presentation.

(4.7) Lemma. Let $(A,\underline{m})$ be a coherent local ring and let E be a finitely presented A-module. Then $pd_A E = n$ iff $Tor_n^A(E,A/\underline{m}) \neq 0$ and $Tor_{n+1}^A(E,A/\underline{m}) = 0$.

Proof. Observe that there is a resolution

$$\dots F_2 \xrightarrow{\phi_2} F_1 \xrightarrow{\phi_1} F_0 \longrightarrow E \longrightarrow 0$$

where each F_i is a free module of finite rank and the matrix ϕ_i has all of its entries in $\underline{m}$. The result follows in just the same way as in the Noetherian case : say, dimension shifting and Nakayama's lemma.

Proof of (4.6) : That $fPD(A) \geq pd_A(A/I) + fPD(A/I)$ follows from A/I, as we observe that A/I is also coherent.

Write $d = pd_A(A/I)$ and let E be an A-module of finite projective dimension and consider a minimal resolution

$$0 \longrightarrow N \longrightarrow F_{d-1} \dots F_0 \longrightarrow E \longrightarrow 0,$$

where *a priori* several of the terms above could be trivial. As $Tor_{d+i}^A(E,A/I) = Tor_i^A(N,A/I)$, $Tor_i^A(N,A/I) = 0$ for $i > 0$.

Consider the spectral sequence

$$E_{p,q}^2 = Tor_p^{A/I}(A/\underline{m},Tor_q^A(N,A/I)) \underset{p}{\longrightarrow} Tor_n^A(A/\underline{m},N)$$

As $E_{p,q}^2 = 0$ if $q > 0$,

$$Tor_p^{A/I}(A/\underline{m},N/IN) = Tor_p^A(A/\underline{m},N).$$

But this isomorphism says that $pd_A N = pd_{A/I}(N/IN)$ and if $pd_A E > d$ we conclude $pd_A E = d + pd_{A/I}(N/IN)$ to get he other inequality $fPD(A) \leq d + fPD(A/I)$.

(4.8) Corollary. Let $(A,\underline{m})$ be a coherent local ring and let I be a finitely generated ideal such that : (a) $pd_A I < \infty$, and (b) the radical of I is $\underline{m}$. Then $fPD(A) < \infty$.

Proof. According to (4.6) (by passing to the ring A/I) we may assume that A is primary ring. Suppose say that

$$0 \longrightarrow F_1 \xrightarrow{\phi} F_0 \longrightarrow E \longrightarrow 0$$

is a minimal resolution of the module E. By McCoy's theorem the minors of rank $\text{rank}(F_1)$ of ϕ have a nontrivial annihilator, impossible unless $F_1 = 0$.

§4.3 Spherical modules.

In this section we take up the question stated at the outset on when is a module E a homomorphic image of a direct sum of copies of a fixed module G. We assume throughout Noetherian conditions. Thus the situation translates on whether the natural map

$$G \otimes_A \text{Hom}_A(G,E) \xrightarrow{\phi_E} E, \quad \phi_E(g \otimes f) = f(g)$$

is surjective. Actually our interest will focus on those modules E admitting resolutions by sums of copies of G. If such resolutions are going to show any form of invariance the condition

i) $\text{Ext}^i_A(G,G) = 0$ for $i > 0$,

should be present. A companion restriction, if the notion of

G-dimension is to be defined, is that the module E be such that

ii) $\quad \mathrm{Ext}^i_A(G,E) = 0 \quad$ for $i > 0$.

This evidently allows for a form of Schanuel's lemma; it will appear quite naturally.

Finally, and in order to attach a divisor to modules that admit finite G-resolutions, we shall need

iii) $\quad \mathrm{Hom}_A(G,G) = A$.

In order to motivate the usefulness of this context we discuss the following result (also proved by T.Gulliksen).

(4.9) Proposition. Let $(A,\underline{m})$ be a local Artinian ring with $\underline{m}^2 = 0$. Let G be a finitely generated A-module that satisfies the conditions

(a) $\quad \mathrm{Hom}_A(G,G) = A$,

(b) $\quad \mathrm{Ext}^1_A(G,G) = 0$.

Then $G \cong A$ or $G \cong I$ (=injective envelope of $\underline{k} = A/\underline{m}$).

Proof. Consider the map $\phi_I : G \otimes \mathrm{Hom}(G,I) \longrightarrow I$. By applying the functor $\mathrm{Hom}(-,I)$ we get

$\mathrm{Hom}(I,I) \longrightarrow \mathrm{Hom}(G \otimes \mathrm{Hom}(G,I),I) \cong \mathrm{Hom}(G,\mathrm{Hom}(G,I),I)) \cong$
$\cong \mathrm{Hom}(G,G)$. But the two ends of this series are isomorphic to A and the combined map is clearly the identity after the appropriate identifications; we conclude that ϕ_I is an isomorphism.

Denote now by $r(E)$ (resp.$\nu(E)$) the dimension of the socle of the module E (resp. minimal number of generators of E). Because $\mathrm{Hom}(-,I)$ is a self-dualizing functor we get from the isomorphism above that $r(G)\cdot\nu(G) = \nu(I) = r(A) = \mu(A)$, the type of the local ring A. In particular this equality says that if A is a Gorenstein ring or more generally the dimension of the socle of A is a prime number then $G \cong A$ or I.

Let us now go back to the conditions of the proposition. Write $s(E)$ for the socle of the module E. In our case $s(A) = \underline{m}$ and $s(G) = \underline{m}G$ as can be easily shown ([15]).

Let

$$0 \longrightarrow L \longrightarrow A^{\nu(G)} \longrightarrow G \longrightarrow 0$$

be a minimal presentation of G. Applying $\mathrm{Hom}(-,G)$ and using (b) we get the exact sequence

$$0 \longrightarrow A \longrightarrow G^{\nu(G)} \longrightarrow \mathrm{Hom}(L,G) \longrightarrow 0.$$

Since $L \subseteq \underline{m}A^{\nu(G)}$ we get that $\mathrm{Hom}(L,G) = \mathrm{Hom}(L,\underline{m}G) = \mathrm{Hom}(L/\underline{m}L, s(G))$. Noting that $L/\underline{m}L = \mathrm{Tor}_1^A(G,\underline{k})$ and taking lengths of the sequence we get

$$\ell(\mathrm{Tor}_1^A(G,\underline{k}))\cdot r(G) + \ell(A) = r(G)\cdot\ell(G).$$

Another way of computing the length of the Tor is via

$$0 \longrightarrow \mathrm{Tor}_1^A(G,\underline{k}) \longrightarrow G \otimes \underline{m} \longrightarrow G \longrightarrow G/\underline{m}G \longrightarrow 0;$$

we get now

$$\ell(\mathrm{Tor}_1^A(G,\underline{k})) = \nu(G) + \nu(G)\cdot r(A) - \ell(G).$$

Taking into account the relations:

$$\ell(A) = r(A) + 1, \quad \text{and}$$

$$\ell(G) = \nu(G) + r(G),$$

we find after an elementary substitution that

$$(r(G)^2 - 1)(\nu(G)^2 - 1) = 0.$$

Example. To show the existence of a module such as G over an Artinian ring that is neither A or I consider the case: $A = K[x,y]$, K a field, $(x,y)^2 = 0$, and I the injective envelope of $A/\underline{m}$; note $I \not\cong A$. Let $B = A[u,v]$, $(u,v)^2 = 0$, and let $G = I\otimes_A B$. G is certainly not isomorphic to the injective envelope of the residue field of B - as the dimension of the socle of G is 2. It is also not isomorphic to B.

Definition. A finitely generated module G over a Noetherian ring A will be called spherical if it satisfies the conditions i) and iii) above. We write Sph(A) for the spherical modules over A.

Remarks. The following observations will be used in the sequel without further ado.

(i) If G is a spherical A-module and I is a rank one projective A-module then $I\otimes G$ is again a spherical module as it is locally isomorphic to G. Conversely, if G and G' are locally isomorphic spherical modules then Hom(G,G') is a rank one projective module I and clearly $I\otimes G = G'$.

(ii) If E is a direct summand of G^n, G spherical, by local cancellation E is isomorphic at each localization to a G^m (m possibly varying with the localization). Conversely, suppose E is locally isomorphic to a G^m; there is a presentation

$$G^n \xrightarrow{\phi} E \longrightarrow 0$$

which, we claim, splits. Assume A local and view ϕ as a matrix with entries in A. As all of its entries cannot lie in $\underline{m}$ by Nakayama's lemma, we easily get a splitting going.

Problem. The determination of Sph(A[t]) in terms of Sph(A) looms a challenging question in view already of the difficulties met in the simpler Pic(A[t]).

§4.4 Elementary properties.

(a) If G is a spherical A-module and $\underline{h}: A \longrightarrow B$ is a flat ring homomorphism then $G \otimes_A B$ is a spherical B-module. (The converse is true if B is faithfully flat.)

Proof. Immediate.

(b) As Hom(G,G) = A for a spherical module G, A and G have the same associated primes by (2.9). Thus if x is a regular element of A it is also G-regular. Applying Hom(G,-) to

$$0 \longrightarrow G \xrightarrow{.x} G \longrightarrow G/xG \longrightarrow 0$$

we get G/xG to be spherical. Conversely, if x is an element in the Jacobson radical of A that is both A-regular and G-regular then the converse holds.

We can also conclude that if G is S-spherical for a local ring A then $\underline{m}$-depth G = depth A. It follows that if A is a Gorenstein local ring then $A \cong G$.

Given a spherical module G the modules satisfying

$$\mathrm{Ext}^i(G,E) = 0 \quad \text{for } i > 0,$$

will play an intriguing role later. For such modules we define a filtration

$$C(G) \subseteq C(G)_0 \subseteq C(G)_\infty \subseteq H(G)$$

as follows

$C(G) = \{ E \mid$ the natural map $\phi_E : G \otimes \mathrm{Hom}(G,E) \longrightarrow E$ is onto $\}$,

$C(G)_0 = \{ E \mid$ with ϕ_E an isomorphism $\}$.

Notice that E belongs in C(G) if it can be written as

$$0 \longrightarrow K \longrightarrow G^n \longrightarrow E \longrightarrow 0.$$

If we apply Hom(G,-) and follow with G⊗(-) we obtain the diagram

$$\begin{array}{ccccccccc} (?) & \longrightarrow & G\otimes\mathrm{Hom}(G,K) & \longrightarrow & G\otimes\mathrm{Hom}(G,G^n) & \longrightarrow & G\otimes\mathrm{Hom}(G,E) & \longrightarrow & G\otimes\mathrm{Ext}(G,K) \\ & & \downarrow & & \downarrow & & \downarrow & & \\ 0 & \longrightarrow & K & \longrightarrow & G^n & \longrightarrow & E & \longrightarrow & 0 \end{array}$$

where the upper sequence begins with an injection and ends with a surjection. If $E \in C(G)_o$ we conclude that $\mathrm{Ext}(G,K) = 0$. Thus $K \in C(G)$ and we also recognize (?) as $\mathrm{Tor}(G,\mathrm{Hom}(G,E))$. Thus $E \in C(G)_o$ implies the existence of a presentation

$$G^m \longrightarrow G^n \longrightarrow E \longrightarrow 0.$$

(c) Schanuel's lemma : Suppose $E \in C(G)_o$ and

(i) $$0 \longrightarrow K \longrightarrow G^m \xrightarrow{\ \phi\ } E \longrightarrow 0$$

(ii) $$0 \longrightarrow L \longrightarrow G^n \xrightarrow{\ \psi\ } E \longrightarrow 0$$

are two G-presentations of E. Then $L \oplus G^m \cong K \oplus G^n$.

Proof. As in the proof of the Schanuel's lemma proper we define $X = \{ (x,y) \in G^m \times G^n \mid \phi(x) = \psi(y)\}$ and have the exact sequences

$$0 \longrightarrow K \longrightarrow X \longrightarrow G^n \longrightarrow 0$$

$$0 \longrightarrow L \longrightarrow X \longrightarrow G^m \longrightarrow 0$$

that split by the preceding remarks.

$C(G)_\infty$ may now be defined as consisting of the modules admitting G-resolutions of arbitrary length. If we now extend the notion of G-resolution by allowing direct summands of G^m's to participate, we define H(G) as consisting of modules with a finite G-resolution. For the modules in H(G) we shall see that the condition $\mathrm{Ext}^i_A(G,-) = 0$, $i > 0$, is automatically satisfied.

(d) Let G, G' be spherical A-modules and x a nonzero divisor in the Jacobson radical of A. If $G/xG \cong G'/xG'$ then $G \cong G'$.

<u>Proof</u>. Apply $\mathrm{Hom}(G,-)$ to

$$0 \longrightarrow G' \xrightarrow{.x} G' \longrightarrow G'/xG' \longrightarrow 0 :$$

$$0 \longrightarrow \mathrm{Hom}(G,G') \longrightarrow \mathrm{Hom}(G,G') \longrightarrow \mathrm{Hom}(G,G'/xG')$$

$$\longrightarrow \mathrm{Ext}^1(G,G') \xrightarrow{.x} \mathrm{Ext}^1(G,G') \longrightarrow \mathrm{Ext}^1(G,G'/xG').$$

Since $\mathrm{Ext}^1_A(G,G'/xG') \cong \mathrm{Ext}^1_{A/(x)}(G/xG,G'/xG')$ and $G/xG \cong G/xG'$ is a spherical $A/(x)$-module, $\mathrm{Ext}^1_A(G,G') = 0$ by Nakayama's lemma. Let now $\phi \in \mathrm{Hom}(G,G')$ be such that it induces the isomorphism $G/xG \cong G'/xG'$; by Nakayama's lemma again ϕ is surjective. Similarly G is a homomorphic image of G'. Thus G and G' are isomorphic by an standard argument.

(e) On the number of isomorphism classes of spherical modules: Let A be a local Macaulay ring; according to (d), the map

$$\mathrm{Sph}(A) \longrightarrow \mathrm{Sph}(A/xA)$$

obtained by reducing modulo a nonzero divisor is injective. In order to estimate (at least whether is finite) the cardinality of $\mathrm{Sph}(A)/\sim$ we assume then that A is an Artinian ring.

Let I denote the injective envelope of $A/\underline{m}$. Then $G \in \mathrm{Sph}(A)$ leads to $G^* = \mathrm{Hom}(G,I) \in \mathrm{Sph}(A)$ also. Indeed

$$\mathrm{Hom}(G^*,G^*) \cong \mathrm{Hom}(G^* \otimes G,I) \cong \mathrm{Hom}(I,I) \cong A.$$

since we saw in (4.9) that $G \otimes G^* \cong I$. As A is Noetherian we recall from [<u>10</u>,Chap.VI] the isomorphism

$$\mathrm{Hom}_A(\mathrm{Ext}^i_A(X,Y),Z) \cong \mathrm{Tor}^A_i(X,\mathrm{Hom}_A(Y,Z))$$

when Z is A-injective and X finitely generated. Making $X = Y = G$ and $Z = I$ we get

$\mathrm{Ext}_A^i(G^*,G^*) \cong \mathrm{Tor}_i^A(G^*,G) \cong \mathrm{Tor}_i^A(G,G^*) \cong (\mathrm{Ext}_A^i(G,G)^*$ for $i > 0$. In this fashion we obtain an involution acting on Sph(A).

Question. By an abuse of notation still write Sph(A) for the isomorphism classes of spherical modules. (a) Is the number of elements in Sph(A), for a local Macaulay ring A, finite? (b) If Sph(A) $\neq$ {A}, is Sph(A) an even number? (It is easy to see that if the residue field of A is finite then Sph(A) is finite.)

The preceding observations also suggest:

(f) If A is a local Macaulay ring admitting a canonical module then $\mathrm{Sph}(A) \longrightarrow \mathrm{Sph}(A/xA)$ is surjective, that is, every element of Sph(A/xA) is liftable. Although we cannot prove this yet, portions of (e) can be extended to higher dimensions. We use an argument of [13,36].

(4.10) Proposition. Let G be an spherical module over the Noetherian ring A and let E be a module of finite injective dimension. Then $\phi_E : G\otimes\mathrm{Hom}(G,E) \longrightarrow E$ is an isomorphism.

Proof. We assume E finitely generated. Again we use the spectral sequences of [10] $\mathrm{Tor}_p(G,\mathrm{Ext}^q(G,E))$ and $\mathrm{Ext}^p(\mathrm{Ext}^q(G,G),E)$ that converge to the same limit. By (2.20) $\mathrm{Ext}^i(G,E) = 0$ for $i > 0$ since depth G = depth A with respect to any ideal. These sequences then yield

$$G\otimes\mathrm{Hom}(G,E) \cong E.$$

(4.11) Corollary. If $G \in \mathrm{Sph}(A)$ and Ω is a canonical module, then $\mathrm{Hom}(G,\Omega) \in \mathrm{Sph}(A)$.

Proof. Firstly, $\mathrm{Hom}(\mathrm{Hom}(G,\Omega),\mathrm{Hom}(G,\Omega)) =$ $\mathrm{Hom}(G\otimes\mathrm{Hom}(G,\Omega),\Omega) = \mathrm{Hom}(\Omega,\Omega) = A$.

Next, to show $\mathrm{Ext}^i(\mathrm{Hom}(G,\Omega),\mathrm{Hom}(G,\Omega)) = 0$ for $i > 0$: Apply $\mathrm{Ext}^j(-,\Omega)$ to this module and get successively

$\mathrm{Ext}^j(\mathrm{Ext}^i(G^*,G^*),\Omega) = \mathrm{Tor}_j(G^*,\mathrm{Ext}^i(G^*,\Omega) =$

$\mathrm{Tor}_j(G^*,\mathrm{Tor}_i(G,\mathrm{Hom}(\Omega,\Omega)) = \mathrm{Tor}_j(G^*,G) = 0, \quad G^* = \mathrm{Hom}(G,\Omega)$.

(g) Reflexivity of spherical modules : Let $G \in \mathrm{Sph}(A)$; if G is reflexive and A is a local Artinian ring,

$$G = \mathrm{Hom}(\mathrm{Hom}(G,A),A)$$

yields the relation

$$r(G) = \nu(\mathrm{Hom}(G,A))\cdot r(A)$$

in the notation of (4.9). Since we have also $r(A) = r(G)\cdot\nu(G)$, we conclude $\nu(\mathrm{Hom}(G,A)) = 1$ and G is A-free.

Assume now that A is a S_2-ring. If G is reflexive we may by the preceding identify G to an ideal of A. As

$$A = \mathrm{Hom}(G,G) = \mathrm{End}(\mathrm{Hom}(\mathrm{Hom}(G,A),A)),$$

for each localization at a prime of height 1 we conclude that $\mathrm{Hom}(G,A)$ is a free A-module by (3.12) and hence G is A-free.

Thus we conclude that if A is a S_2-ring then G is reflexive iff it is isomorphic to a divisorial ideal that is principal at the primes of $P(A)$.

(4.12) Corollary. If A is an UFD then $\mathrm{Sph}(A) = \{A\}$.

This could be viewed as a generalization of (2.39).

§4.5 Resolutions and Divisors.

Given a spherical module G, an Euler characteristic is

attached to the modules in H(G) and a divisor to the modules in H(G) with torsion; this subset will be denoted by $H(G)^o$ and the associated divisor runs parallel to that of Chapter 3.

Assume A to be a connected ring. For any module E which is a direct summand of a G^n we may attach a unique rank in any of the two ways:

i) As $\mathrm{Hom}(G^n,G)$ is free of rank n the rank of E is the rank of the projective module $\mathrm{Hom}(E,G)$.

ii) $E \oplus F = G^n$ implies that at each localization E is isomorphic to a direct sum of r copies of G - local cancellation ; that r is unique and agrees with the rank in i).

(a) Euler <u>characteristic</u> :

Let E be a module in H(G). Let

$$0 \longrightarrow G^{r_n} \xrightarrow{\phi_n} \cdots \longrightarrow G^{r_1} \xrightarrow{\phi_1} G^{r_0} \longrightarrow E \longrightarrow 0$$

be a finite G-resolution.

<u>Definition</u>. $\chi_G(E) = \Sigma(-1)^i r_i$.

This integer is not dependent on the resolution and its nullity is related to the existence - just as in (2.26) - of nonzero divisors of A in the annihilator of E. We show this fact in two different ways.

Because $\mathrm{Hom}(G,G) = A$, we may view the resolution as a G-matricial complex and apply (2.32) :

i) $I(\phi_k)$-depth $G \geq k$;

ii) $\mathrm{rank}(\phi_k) + \mathrm{rank}(\phi_{k-1}) = r_k$.

As every regular A-sequence is also a regular G-sequence,

if we apply Hom(G,-) to this sequence we obtain a free complex that is also exact. If we now tensor this last complex by G we obtain the original complex and

$$\phi_E : G\otimes \mathrm{Hom}(G,E) \longrightarrow E$$

is an isomorphism. In particular the hypothesis that the modules in H(G) satisfy the condition in $C(G)_o$ is not needed. This shows that the integer $\chi_G(E)$ equals the usual Euler characteristic of Hom(G,E) which is a module of finite projective dimension. Note that $\mathrm{Tor}_i(G,\mathrm{Hom}(G,E)) = 0$ for $i > 0$, a remark to be used later.

(4.13) Corollary. The functors Hom(G,-) and $G\otimes(-)$ are inverse equivalences of the categories H(G) and H(A).

(4.14) Corollary. $\chi_G(E) \geq 0$; $\chi_G(E) = 0$ iff ann(E) contains a nonzero divisor of A.

Similarly we could define a G-dimension for the modules in H(G) : assume A to be a local ring and define G-dim(E) as the last integer for which $\mathrm{Ext}^r(E,G) \neq 0$. In this case however nothing new is obtained as this number is just $\underline{m}$-depth A - $\underline{m}$-depth E.

Another approach is to use the Schanuel's lemma of §4.3c together with the techniques of local cancellation and develop a theory of dimension and minimal resolutions.

Remark. Without much difficulty one can see that if in a short exact sequence two modules are in H(G) then so is the third one.

(4.15) Remark. Let G,G' be two spherical modules. The

evidence seems to point that $H(G) \cap H(G')$ is rather sparse if G and G' are not locally isomorphic. Suppose A is a local ring and let E be a module with $G\text{-dim}(E) = 1$; if $E \in H(G')$ we must also have that $G'\text{-dim}(E) = 1$, by an earlier remark. Let

$$0 \longrightarrow G_1 \xrightarrow{\ \phi\ } G_0 \longrightarrow E \longrightarrow 0$$

be a minimal G-resolution of E - i.e. the entries of ϕ are in the maximal ideal of A. We obtain

$$0 \longrightarrow \mathrm{Hom}(G',G_1) \longrightarrow \mathrm{Hom}(G',G_0) \longrightarrow \mathrm{Hom}(G',E) \longrightarrow$$
$$\longrightarrow \mathrm{Ext}^1(G',G_1) \xrightarrow{\ \phi\ } \mathrm{Ext}^1(G',G_0) \longrightarrow \mathrm{Ext}^1(G',E) = 0.$$

By Nakayama's lemma we then have $\mathrm{Ext}^1(G',G) = 0$. Tensor now the sequence above with G' to get the diagram

$$\begin{array}{ccccccccc}
0 & \longrightarrow & K_1 & \longrightarrow & K_0 & \longrightarrow & 0 & & \\
 & & \downarrow & & \downarrow & & & & \\
0 & \longrightarrow & G'\otimes\mathrm{Hom}(G',G_1) & \longrightarrow & G'\otimes\mathrm{Hom}(G',G_0) & \longrightarrow & G'\otimes\mathrm{Hom}(G',E) & \longrightarrow & 0 \\
 & & \downarrow & & \downarrow & & \wr\downarrow & & \\
0 & \longrightarrow & G_1 & \longrightarrow & G_0 & \longrightarrow & E & \longrightarrow & 0
\end{array}$$

where the middle horizontal sequence is exact as the module $\mathrm{Tor}_1(G',\mathrm{Hom}(G',E)) = 0$. From the minimality of the resolution we also conclude that ϕ_G is surjective. K_1 and K_0 on the other hand are direct sums of the kernel of ϕ_G and the map between them is really the 'matrix' ϕ. By Nakayama's lemma again we get that ϕ_G is an isomorphism. Counting minimal numbers of generators we have $\nu(G')\cdot\nu(\mathrm{Hom}(G',G)) = \nu(G)$. Taking into account the symmetrical relation we conclude $\nu(\mathrm{Hom}(G',G)) = 1$. This last module being faithful it is isomorphic to A and $G \cong G'$.

(4.16) Remark. A statement more general than §4.4b is the

following : Let G be a spherical A-module and let I be a Gorenstein ideal, that is, an ideal satisfying

i) grade I = pd A/I = r, and

ii) $\mathrm{Ext}_A^r(A/I,A) = A/I$.

Then $G/IG \in \mathrm{Sph}(A/I)$.

Proof. We may assume that A is a local ring. Consider the following convergent spectral sequence

$$\mathrm{Ext}^p_{A/I}(G/IG,\mathrm{Ext}^q_A(A/I,G)) \longrightarrow \mathrm{Ext}^n_A(G/IG,G).$$

Since every regular A-sequence is G-regular, $\mathrm{Ext}^q_A(A/I,G) = 0$ for $q < r$ and $\mathrm{Ext}^r_A(A/I,G) \cong \mathrm{Ext}^r_A(A/I,A)\otimes G \cong G/IG$. Thus we get the isomorphism

$$\mathrm{Ext}^p_{A/I}(G/IG,G/IG) \cong \mathrm{Ext}^{p+r}_A(G/IG,G).$$

Let now

$$0 \longrightarrow F_r \xrightarrow{\ \phi\ } F_{r-1} \cdots F_0 \longrightarrow A/I \longrightarrow 0$$

be a minimal projective resolution of A/I. The condition ii) makes F_r of rank 1 and $\phi(1) = (a_1,\dots,a_n) \in F_{r-1}$ with the a_i's a minimal generating set for the ideal I. Tensor this sequence with G to get a G-resolution of length r for G/IG and consequently $\mathrm{Ext}^{p+r}_A(G/IG,G) = 0$ for $p > 0$. We also get $\mathrm{Ext}^r_A(G/IG,G) \cong A/I$ to complete the proof.

(b) Divisors :

Now we attach an invertible ideal $\underline{d}(E)$ to any torsion module $E \in H(G)$.

i) Let E be a torsion module in $C(G)_o$; consider a presentation

$$G^m \xrightarrow{\phi} G^n \longrightarrow E \longrightarrow 0.$$

Define $\underline{d}(E) = (F(\phi)^{-1})^{-1}$, where $F(\phi)$ denotes the ideal generated by the minors of order n of the matrix. By the Schanuel's lemma of §4.4c we get that $\underline{d}(E)$ does not depend on the G - presentation.

ii) Suppose now $E \in H(G)$:

$$0 \longrightarrow G_n \; \cdots \; G_0 \longrightarrow E \longrightarrow 0$$

(where G_i is a direct summand of a G^m). Applying Hom(G,-) we obtain the module $\mathrm{Hom}(G,E) \in H(A)$ and $\underline{d}(E)$ is the divisor introduced in Chapter 3 of the module Hom(G,E). Thus $\underline{d}(E)$ is an invertible ideal.

iii) Suppose now $0 \neq E \in H(G) \cap H(G')$ for distinct spherical modules. As we remarked, it is plausible that $H(G) = H(G')$ at least in the local case. Nevertheless to obtain the equality of the two reflexive ideals $\underline{d}_G(E)$ and $\underline{d}_{G'}(E)$ it is enough to compare them at the localizations $A_{\underline{p}}$, $\underline{p} \in P(A)$. But in this case we use (4.15) that says $E_{\underline{p}} = 0$ or $G_{\underline{p}} \cong G'_{\underline{p}}$.

To sum up :

(4.17) <u>Theorem</u>. Let Sph(A) denote the set of spherical modules over A and let E be a module in $H(G)^o$. Then $\underline{d}_G(E)$ is an invertible ideal and does not depend on G. Restricted to a fixed H(G) this divisor is additive.

Chapter 5

I-divisors

A divisor is attached to finitely generated modules over a Noetherian ring that does not quite arise in the manner of Chapter 4. It is obtained by mimicking the construction of the standard Fitting's divisors but using injective resolutions.

Thus if A is a Noetherian ring and E is a finitely generated module we consider a resolution

$$0 \longrightarrow E \longrightarrow I^0 \xrightarrow{\phi} I^1$$

with I^0 and I^1 injective modules. We assume further that, for each prime $\underline{p}$, $I(A/\underline{p})$ appears finitely often only in the indecomposable representations of I^0 and I^1 - and this can always be arranged by [3]. ϕ may then be viewed as a matrix and the determinantal ideals formed much as done earlier. To overcome several technical difficulties - as these ideals live in a product of $\underline{p}$-adic completions - shall be assumed that primes of height one and grade one are the same. The gluing can then be carried out and divisorial ideals in A obtained.

After a summary examination of its properties this divisor is applied to the category of modules of finite injective dimension where it turns out to be invertible.

§5.1 Construction.

Let A be a Noetherian ring of type S_2 - the grade restriction above. This is a minor constraint as the modules of finite injective dimension are effectively defined over these rings.

Let E be a finitely generated torsion module and consider an injective presentation as above. In the set Ass(E) of associated primes of E consider those primes $\underline{p}_1,\ldots,\underline{p}_n$ of height one. If there is no such prime in Ass(E) we put I-divisor of E = div(E) = A. Assume this is not the case and let S be the multiplicative set $\cap\,(A\setminus \underline{p})$. Localize the presentation at S to get an injective presentation of E_S over the one-dimensional semi-local ring A_S. The injective modules $(I^0)_S$ and $(I^1)_S$ are direct sums of the $I(A/\underline{p}_i)$'s only. Change notation and assume for the moment A is A_S. The resolution takes a more explicit form :

$$0 \longrightarrow E \longrightarrow \bigoplus_i I(A/\underline{p}_i)^{r_i} \xrightarrow{\ \phi\ } \bigoplus_i I(A/\underline{p}_i)^{s_i}.$$

We recall that ([27]) :

i) $\mathrm{Hom}_A(I(A/\underline{p}),I(A/\underline{p})) = \hat{A}_{\underline{p}}$ (where '^' over a module denotes completion with respect to the adic topology of the corresponding maximal ideal).

ii) $\mathrm{Hom}_A(I(A/\underline{p}),I(A/\underline{q})) = 0$ for incomparable primes $\underline{p},\underline{q}$.

Altogether these remarks lead to a view of ϕ as a matrix

$$\begin{bmatrix} B_1 & & 0 \\ & \ddots & \\ 0 & & B_n \end{bmatrix}$$

where B_i is an $r_i \times s_i$ block with entries in $\hat{A}_{\underline{p}_i}$. The ideal $D_i(E)$ of $\hat{A}_{\underline{p}_i}$ generated by the minors of order r_i of B_i is easily seen to be a $\underline{p}_i\hat{A}_{\underline{p}_i}$-primary: Localize E at $\underline{p}_i$ - and $E_{\underline{p}_i}$ becomes a module of finite length - and apply the dualizing functor $(-)^* = \mathrm{Hom}_{\hat{A}_{\underline{p}_i}}(-,I(A/\underline{p}_i))$ to obtain the exact sequence

$$\hat{A}_{\underline{p}_i}^{s_i} \xrightarrow{{}^tB_i} \hat{A}_{\underline{p}_i}^{r_i} \longrightarrow E^* \longrightarrow 0 \quad (t=\text{transpose})$$

and the ideal $D_i(E)$ just defined is the 0-th Fitting ideal of E^*. In particular it is independent of the chosen injective presentation.

There is a unique primary ideal $\underline{q}_i$ of A such that $\underline{q}_i\hat{A}_{\underline{p}_i} = D_i(E)$. It then follows that we may find an ideal $\underline{q}$ in A such that $(\underline{q}A_{\underline{p}_i})\hat{} = D_i(E)$ for each of the $\underline{p}_i$'s.

We now go back to the original situation before the localization at S. We pick an ideal Q in A with $\underline{p}_1,\ldots,\underline{p}_n$ for its only associated primes and with $QA_S = \underline{q}$.

(5.1) Definition. $\operatorname{div}(E) = (Q^{-1})^{-1} = D(Q)$.

Example. Let k be a field and let t be an indeterminate. Let $B = k[[t^3,t^4]] \hookrightarrow A = k[[t^3,t^4,t^5]] \hookrightarrow k[[t]]$. Notice that B is a Gorenstein ring and that $\Omega = \operatorname{Hom}_B(B,A)$ is a canonical module for A. It is clear that $\Omega = (t^3,t^4)$. Let E be the module $\Omega/t^3\Omega$. E has a presentation

$$0 \longrightarrow L \longrightarrow A^2 \longrightarrow E \longrightarrow 0$$

where L has for generators the elements $(t^3,0)$, $(0,t^3)$, (t^4,t^5), (t^5,t^4) and (t^6,t^5). Thus the Fitting divisor of E, $\underline{d}(E) = t^3(t^3,t^4,t^5)$. As for its I-divisor :

$$0 \longrightarrow \Omega \xrightarrow{.t^3} \Omega \longrightarrow E \longrightarrow 0$$

leads to

$$0 \longrightarrow A \xrightarrow{.t^3} A \longrightarrow \operatorname{Ext}_A^1(E,\Omega) \longrightarrow 0.$$

Finally, as $\operatorname{Ext}_A^1(E,\Omega) = \operatorname{Hom}_A(E,I(A/\underline{m})$, we conclude that

$$\underline{d}(\operatorname{Ext}_A^1(E,\Omega)) = t^3A = \operatorname{div}(E).$$

§5.2 Euler characteristics of Inj(A).

For a Noetherian ring A we develop a theory of Euler characteristics for Inj(A), the category of finitely generated modules of finite injective dimension as considered in [36,38]. In the next section a divisor theory for the subcategory Inj $\mathrm{Inj}(A)^{o}$ consisting of the modules in Inj(A) with torsion will be discussed. Both depend crucially on in some partial duality established in [30] between Inj(A) and the modules of finite projective dimension.

Without loss of generality let A be a connected ring. Let E be a finitely generated module of finite injective dimension, and write

$$0 \longrightarrow E \longrightarrow I^0 \longrightarrow I^1 \ldots \; I^n \longrightarrow 0$$

for a minimal injective resolution of E. For a prime ideal $\underline{p}$ with depth $A_{\underline{p}} = r$ write

$$\chi(\underline{p};E) = \sum_i (-1)^{r-i}\mu_i(\underline{p};E).$$

(5.2) Theorem. $\chi(\underline{p};E) = \chi(E)$ is a non-negative integer that does not depend on $\underline{p}$. $\chi(E) = 0$ iff the annihilator of E contains a regular element.

That χ is then an Euler characteristic on Inj(A) follows from the meaning of the μ_i's.

Proof. Because of the connectedness of A we may assume that A is a local ring of maximal ideal $\underline{m}$. Let r = depth A: let $\hat{A}$ denote the $\underline{m}$-adic completion of A. As $I_A(A/\underline{m}) = I_{\hat{A}}(A/\underline{m})$, in calculating $\chi(\underline{m};E)$ we may as well assume that A is a complete local ring.

The setting is now ready for an application of ([30]) :

(5.3) Theorem. Let A be a complete local ring of depth r and E be a finitely generated module of finite injective dimension. Then the module $M = \mathrm{Ext}_A^r(I(A/\underline{m}),E)$ has the following properties : i) M is a finitely generated module of finite projective dimension; ii) $\mathrm{Supp}(M) = \mathrm{Supp}(E)$; iii) $\mathrm{Ext}_A^i(M,A) = \mathrm{Hom}(H_{\underline{m}}^{r-i}(E), I(A/\underline{m}))$, where $H_{\underline{m}}^*(-)$ stand for the local cohomology groups.

Proof of (5.2) : (a) It will be first shown that $\mu_i(E) = \beta_{r-i}(M)$ where $\beta_j(M) = \dim_{\underline{k}}(\mathrm{Tor}_j^A(\underline{k},M))$ = rank of the j-th term in a minimal free resolution of M. In this case $\chi(\underline{p};E)$ would equal the Euler characteristic defined in §2.5; the statement on the positivity and relation to the regularity of the annihilator would follow from ii) above. To show the equality of the μ_i and β_{r-i} consider the convergent spectral sequences with the same limit

$$\mathrm{Tor}_p(\underline{k},\mathrm{Ext}^q(I,E)) \quad \text{and} \quad \mathrm{Ext}^p(\mathrm{Ext}^q(\underline{k},I),E) \quad (I = I(A/\underline{m})).$$

In the proof of (5.3) it emerges that $\mathrm{Ext}^i(E,I) = 0$ for $i < r$. Consequently we have $\mathrm{Tor}_{r-i}(\underline{k},M) = \mathrm{Ext}^i(\underline{k},E)$.

(b) This step consists in showing that if $\underline{p}$ is any prime ideal then $\chi(\underline{p};E) = \chi(\underline{m};E)$. In this we may assume that $\chi(\underline{m};E) > 0$, that is, E is a faithful A-module. From (2.34) this makes A a Macaulay ring. If A admitted a canonical Ω - as is the case for $\hat{A}$ - we would have

$$\Omega\otimes\mathrm{Hom}(\Omega,E) \xrightarrow{\sim} E \quad (4.10) \quad \text{and}$$

$$\mathrm{Ext}^i(\Omega,E) = 0,\ i > 0 \quad (2.20)$$

and could then begin a resolution

$$0 \longrightarrow L \longrightarrow \Omega_0 \longrightarrow E \longrightarrow 0$$

with Ω_0 a direct sum of copies of Ω and L another module of finite injective dimension; also, $\underline{m}$-depth L $\geq$ inf{ $\underline{m}$-depth Ω, $\underline{m}$-depth E+1 }. We could then fashion a resolution

$$0 \longrightarrow K \longrightarrow \Omega_{r-1} \cdots \Omega_0 \longrightarrow E \longrightarrow 0$$

with Ω_i = direct sum of Ω's and $\underline{m}$-depth K = r. K would then be isomorphic to some Ω^n and the constancy of $\chi(\underline{p};E)$ would follow much as in §4.5.

To avoid the use of the canonical module we reason as follows : Let $\underline{p}_o$ be a minimal prime ideal of A contained in $\underline{p}$. Let also $\underline{q}_o$ be a minimal prime of Λ lying above $\underline{p}_o$.

(5.4) Lemma.([16]) Let $(A,\underline{m}) \longrightarrow (B,\underline{n})$ be a flat local homomorphism of local Noetherian rings and let $B/\underline{m}B$ be a Macaulay ring of type $r_B(B/\underline{m}B) = r$. Let E be a (f.g.) Macaulay A-module of type $r_\Lambda(E) = s$. Then $E \otimes B$ is a Macaulay B-module and $r_B(E \otimes B) = r \cdot s$.

If we use the 0-dimensional case of this lemma we get

$\chi(\underline{m};E) = \chi(\underline{m};E) = \chi(\underline{q}_o;E) = \chi(\underline{p}_o;E)$ and finally let

$\underline{p}$ play the role of $\underline{m}$.

§5.3 Divisors on Inj(A)^o.

The construction of div(-) in §5.1 applied to the torsion modules in Inj(A) yields a divisorial ideal which will be shown to be invertible. Notice that the hypothesis that Λ be of type S2 is fulfilled as Bass's conjecture is resolved for rings of local depth one.

We discuss first the change of the injective dimension of a module after a flat change of rings $\underline{h} : A \longrightarrow B$. Next, we show that under mild conditions on the fibers of the morphism $\underline{h}$ the divisor introduced in §5.1 behaves properly. Additional facts may be found in [12,37].

Let A be a Noetherian ring and let E be an A-module of finite injective dimension. Now let $\underline{h} : A \longrightarrow B$ be a flat homomorphism of rings. We are interested in the conditions that make $B\otimes_A E$ a B-module of finite injective dimension. In this generality simple examples show the naturality of the following conditions : (a) The Krull dimensions of A and B are finite; (b) The fibers of $\underline{h}$, i.e. the rings $B\otimes_A k(\underline{p})$ are Gorenstein rings.

(5.5) Theorem. $Id_B(B\otimes_A E) < \infty$.

Proof. Let

$$0 \longrightarrow E \longrightarrow I^0 \longrightarrow I^1 \ldots I^n \longrightarrow 0$$

be an injective resolution of E. Since the Krull dimension of a ring bounds the (finite) injective dimension of modules ([4]) and the I^i are direct sums of indecomposable injectives, we may assume that $E = I(A/\underline{p})$. In considering $B\otimes_A E$ we notice that $B\otimes_A E = B_{\underline{p}}\otimes_{A_{\underline{p}}} E$. We may therefore assume that the rings in question are local rings : $(A,\underline{m})$, (B,M). If $\underline{m}B = B$ we get $0 = B\otimes_A I(A/\underline{m})$, from the structure of $I(A/\underline{m})$. The proof will now follow by induction on the dimension of the fiber $B/\underline{m}B$.

(i) $\dim B/\underline{m}B = 0$: We claim that in this case $B\otimes E$ is the injective envelope of B/M. First notice that $B\otimes E$ is an essential extension of $B/\underline{m}B$, as $\mathrm{Hom}_B(B/\underline{m}B, B\otimes E) = B\otimes \mathrm{Hom}(A/\underline{m}, E)$. As $B/\underline{m}B$ is a Gorenstein ring of dimension 0, it is an essential extension of B/M. We may then write

$$0 \longrightarrow B\otimes E \longrightarrow I_B(B/M).$$

To show equality above it is enough to verify that $\mathrm{Hom}_B(B/\underline{m}^n B,)$ takes the same value on the two modules for each n. This reduces the question to the consideration of Artinian rings, i.e. that $\mathrm{Ext}^1_B(B/\underline{m}B, B\otimes E) = 0$ implies $\mathrm{Ext}^1_B(B/M, B\otimes E) = 0$. Suppose we want to complete the diagram

$$\begin{array}{ccc} B & & \\ \cup & \searrow & \\ M & \xrightarrow{\ \phi\ } & B\otimes E \end{array} ;$$

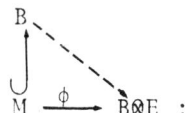

the restriction of ϕ to $\underline{m}B$ can be lifted to a map $\psi: B \rightarrow B\otimes E$. To prove that ϕ may be lifted it suffices to show that $\phi - \psi$ is liftable. A simple induction on $M^n + \underline{m}B$ leads to the desired result.

(ii) $\dim B/\underline{m}B > 0$: Suppose that $\mathrm{id}_B(B\otimes E) = \infty$. Among all prime ideals of B lying above $\underline{m}$ pick a minimal one such that $\mathrm{id}_{B_Q}(B\otimes E)_Q = \infty$. Change the notation and assume $Q = M$. Let x be a nonzero divisor in the radical of $B/\underline{m}B$; then x is a nonzero divisor in B and B/xB is A-flat. Thus the sequence

$$\text{(S)} \qquad 0 \longrightarrow B\otimes E \xrightarrow{\ .x\ } B\otimes E \longrightarrow (B/xB)\otimes E \longrightarrow 0$$

is exact and by the induction hypothesis $\mathrm{id}_{B/xB}(B/xB\otimes E) < \infty$. By the change of dimension theorem $\mathrm{id}_B(B/xB\otimes E) < \infty$ also. Thus for large n (say $\geq 1 + \dim B$) we have

$$\mathrm{Ext}^n_B(-, B\otimes E) \xrightarrow[\sim]{\ .x\ } \mathrm{Ext}^n_B(-, B\otimes E)$$

and in particular $\mathrm{Ext}_B^n(B/M, B\otimes E) = 0$. Suppose for some prime ideal P of B, $\mathrm{Ext}_B^n(B/P, B\otimes E) \neq 0$; pick P maximal and let $a \in M \setminus P$. The sequence

$$0 \longrightarrow B/P \xrightarrow{\cdot a} B/P \longrightarrow B/(P,a) \longrightarrow 0$$

yields

$$\mathrm{Ext}_B^n(B/P, B\otimes E) \overset{\cdot a}{\cong} \mathrm{Ext}_B^n(B/P, B\otimes E).$$

Thus $\mathrm{Ext}_B^n(B/P, B\otimes E)$ is divisible uniquely by every element in $\underline{m} \setminus (P \cap A)$ and should be left unchanged if we localize at the multiplicative set $S = A \setminus (P \cap A)$. But $B\otimes E\otimes A_S = 0$.

The simplest example where (5.3) applies is $B = A[t]$. More generally, if $B = A[t]/I$ is A-flat then I is a projective ideal of A[t] and the condition on the fibers is automatically satisfied. Suppose $B = A[x,y]/I$ is A-flat with Gorenstein fibers and let us examine quickly what this portends for I.

We first remark that J the ideal generated by the coefficients of the polynomials in I is generated by one idempotent ([40]). Without sacrifice of generality we may assume $J = A$. Suppose A to be local with maximal ideal $\underline{m}$. Tensoring the sequence

$$0 \longrightarrow I \longrightarrow A[x,y] \longrightarrow B \longrightarrow 0$$

with $A/\underline{m}$ we get

$$0 \longrightarrow I/\underline{m}I \longrightarrow A/\underline{m}[x,y] \longrightarrow B/\underline{m}B \longrightarrow 0.$$

If $B/\underline{m}B$ is to be a Gorenstein ring then either (a) $I/\underline{m}I$ is principal, or (b) $I/\underline{m}I$ is generated by two elements without a common factor. We can easily conclude

(5.6) Corollary. I is an ideal of projective dimension ≤ 1 and $B = B_1 \times B_2$ where $B_1 = A[x,y]/I_1$, I_1 a projective ideal and B_2 quasi-finite (i.e. with finite fibers).

To have an idea of more general flat extensions of finite type of a ring A we look at the following extension of the Hilbert's syzygies theorem.

Let A be a commutative ring - not necessarily Noetherian - and let E be a module of finite presentation over $B = A[\underline{t}] = A[t_1,\ldots,t_n]$ - or more generally a flat A-algebra of finite presentation, with regular fibers.

(5.7) Proposition. If E is a flat A-module $pd_B E \leq n$.

Proof. The key element is the fact that E - under these circumstances - admits an infinite finite presentation in accordance with [14, Prop.11.3.9.1] :

$$0 \longrightarrow L \longrightarrow P_{n-1} \cdots P_0 \longrightarrow E \longrightarrow 0$$

with the P_i's finitely generated projective B-modules and L finitely presented. L is also a flat A-module. Now we show that L is a projective B-module. We may localize at a prime P of B; let $\underline{p} = P \cap A$. By Hilbert's syzygies theorem $(L/\underline{p}L)_P$ is a projective $B_P/\underline{p}B_P$-module. To conclude write

$$0 \longrightarrow K \longrightarrow F \longrightarrow L_P \longrightarrow 0$$

a minimal free presentation of L_P. Tensor with $k(\underline{p})$ to get

$$0 \longrightarrow K \otimes k(\underline{p}) \longrightarrow F \otimes k(\underline{p}) \longrightarrow L_P \otimes k(\underline{p}) \longrightarrow 0$$

still exact since L is A-flat. By Nakayama's lemma $K = 0$.

Now we return to the consideration of the finitely generated torsion modules over a ring A and their divisors. Suppose E lies in $\mathrm{Inj}(A)^{o}$ and let $\underline{h} : A \longrightarrow B$ be a faithfully flat homomorphism of rings with finite Krull dimension and Gorenstein fibers at the maximal ideals (The case to keep in mind is that of a local ring $(A,\underline{m})$ and its completion).

(5.8) Theorem. $B\otimes E \in \mathrm{Inj}(B)$ and $\mathrm{div}_B(B\otimes E) = \mathrm{div}_A(E)\cdot B$.

Proof. We may assume that $\underline{h}$ is a local homomorphism of local rings. That $E \in \mathrm{Inj}(A)$ can be interpreted for A local as $\mathrm{Ext}^n_A(A/\underline{m},E) = 0$ for large n. The proof of (5.5) leads, via Nakayama's lemma, to the conclusion $B\otimes E \in \mathrm{Inj}(B)$.

Let Q be a prime ideal of B, of height 1 and minimal over the annihilator $J\cdot B$ of $B\otimes E$, where $J = \mathrm{ann}_A E$. Let $P = \underline{h}^{-1}(Q)$. Then P is also a height 1 prime minimal over J. Complete A_P and B_Q with respect to the topologies defined by the maximal ideals in each ring. As $\hat{A}_P$ admits a canonical module Ω there is a resolution

$$0 \longrightarrow \Omega^n \xrightarrow{\ \phi\ } \Omega^n \longrightarrow \hat{E} \longrightarrow 0.$$

Now it follows - by Nakayama's lemma - that $\Omega\otimes\hat{B}_Q$ is a module of injective dimension one and also of rank one as

$$0 \longrightarrow (\Omega\otimes\hat{B}_Q)^n \xrightarrow{\ \phi\otimes 1\ } (\Omega\otimes\hat{B}_Q)^n \longrightarrow (E\otimes\hat{B}_Q) \longrightarrow 0$$

is exact, $E\otimes\hat{B}_Q \in \mathrm{Inj}(\hat{B}_Q)$ and the entries of $\phi\otimes 1$ lie in the maximal ideal of $\hat{B}_Q$. Thus $\Omega\otimes\hat{B}_Q$ is also a canonical module for $\hat{B}_Q$. From this it follows that $\mathrm{div}(E\otimes B)\hat{} = \det(\phi\otimes 1)\hat{B}_Q = \det\phi\cdot\hat{B}_Q$.

(5.9) Theorem. If $E \in \mathrm{Inj}(A)^{o}$, $\mathrm{div}(E)$ is an invertible ideal.

Since the definition of div(E) 'commutes' with localization, A may be assumed local; using now (5.8) we may take A to be complete with respect to the $\underline{m}$-adic topology.

Proof. Let M be the module of (5.3). We show that the Fitting divisor $\underline{d}(M) = \mathrm{div}(E)$. Let $\underline{p}_1,\ldots,\underline{p}_n$ be the primes of height one in $\mathrm{Supp}(E) = \mathrm{Supp}(M)$. To show the equality above suffices to check the localizations at these primes.

Let R be a d-dimensional regular local ring mapping onto A. As the Krull dimension of E is depth A - 1 by (2.34) there is an isomorphism given by local duality

$$\mathrm{Ext}_A^1(M,A) \cong \mathrm{Ext}_R^{d-r+1}(E,R)$$

with r = depth A. Let P be one of the $\underline{p}_i$'s and Q a prime of R lying above P. Localizing this isomorphism at Q we obtain

$$\mathrm{Ext}^1_{A_P}(M_P,A_P) \cong \mathrm{Ext}^{d-r+1}_{R_Q}(E_P,R_Q).$$

Notice that d-r+1 = Krull dimension $R_Q = t$ and that E_P is an R_Q-module of finite length. The second module can also be written, from the spectral sequence of [10, XVI.5] as

$$\mathrm{Ext}^1_{A_P}(E_P,\mathrm{Ext}^{t-1}_{R_Q}(A_P,R_Q)).$$

Since $\mathrm{Ext}^{t-1}_{R_Q}(A_P,R_Q)$ is the canonical module for A_P, we conclude that $\mathrm{div}(E)_P$ is the 0-th Fitting ideal of the module $\mathrm{Ext}^1_{A_P}(M_Q,A_P)$. But it is clear that $M_Q = M_P$ and $\mathrm{Ext}^1_{A_P}(M_Q,A_P)$ have the same invariants as $\mathrm{pd}_{A_P} M_P = 1$ and M_P is a torsion module.

We may now state an interesting application. Let E be a finitely generated module of finite injective dimension over the local ring A. Let J = ann(E).

(5.10) Corollary. If grade J = 1, then A is a Macaulay ring.

Proof. Write Ad = div(E) and let $\underline{p}_1, \ldots, \underline{p}_n$ be the prime ideals of height one above J. By the properties of div(-) the ideal Ad shares all of these primes and has no other associated primes. Thus

$$\text{Krull dim } (A/dA) = \sup\{ \text{ Krull dim } (A/\underline{p}_i) \}.$$

But if r = depth A, all of those modules $A/\underline{p}_i$ have Krull dimension r-1. Since the Krull dimension of A/dA is the dimension of A less one we get r = dim A.

Bibliography

[1] M.Auslander, Coherent Functors, *in* Proc.Conf.Categorical Algebra, Springer-Verlag, 1965, 189-231

[2] - - and D.Buchsbaum, Homological dimension in local rings, Trans.Amer.Math.Soc. 85 (1957), 390-405.

[3] H.Bass, On the ubiquity of Gorenstein rings, Math. Zeitschr. 82 (1963), 8-28.

[4] - - , Injective dimension in Noetherian rings, Trans. Amer.Math.Soc. 102 (1962), 18-29.

[5] - - , A.Heller and R.G.Swan, The Whitehead group of a polynomial extension, Publ.Math.IHES 22, Paris, 1964.

[6] N.Bourbaki, Algèbre Commutative, Chaps. I - VII, Hermann, Paris, 1960-65.

[7] D.Buchsbaum and D.Eisenbud, What makes a complex exact ?, J.Algebra 25 (1973), 259-268.

[8] - - - , Gorenstein ideals of height three, Preprint.

[9] L.Burch, On ideals of finite homological dimension in local rings, Proc.Camb.Phil.Soc. 64 (1968), 941-948.

[10] H.Cartan and S.Eilenberg, Homological Algebra, Princeton University Press, Princeton, 1956.

[11] H.-B.Foxby, On the μ^i in a minimal injective resolution, Math.Scand. 20 (1971), 175-186.

[12] - - , Injective modules under flat base change, Preprint.

[13] - - , On Gorenstein modules and related modules, Math. Scand. 31 (1972), 267-281.

[14] A.Grothendieck, Eléments de Géométrie Algébrique, Publ. Math.IHES 32, Paris, 1965.

[15] T.H.Gulliksen, On the length of faithful modules over Artinian local rings, Math.Scand. 31 (1972), 78-82.

[16] J.Herzog and E.Kunz, Der kanonische Modul eines Cohen - Macaulay Rings, Lectures Notes in Mathematics 238, Springer-Verlag, Berlin, 1971.

[17] M.Hochster, Deep local rings, Preprint.

[18] - - and J.L.Roberts, Actions of reductive groups on regular rings and Cohen - Macaulay rings, Bull.Amer.Math. Soc. 80 (1974), 281-284.

[19] C.U.Jensen, On the vanishing of $\varprojlim^{(n)}$, J.Algebra 15 (1970), 151-166.

[20] I.Kaplansky, Commutative Algebra, Allyn and Bacon, Boston, 1971.

[21] - - , R-sequences and homological dimension, Nagoya Math.J. 20 (1962), 195-200.

[22] H.Krämer, Einege Anwendungen der G-Function von MacRae, Arch.Math. 22 (1972), 479-490.

[23] D.Lazar, Autour de la platitude, Bull.Soc.Math. France 97 (1969), 81-128.

[24] G.Levin and W.V.Vasconcelos, Homological dimensions and Macaulay rings, Pacific J.Math. 25 (1968), 315-323.

[25] J.Lipman, On the Jacobian ideal of the module of differentials, Proc.Amer.Math.Soc. 21 (1969), 422-426.

[26] R.MacRae, On an application of the Fitting invariants, J.Algebra 2 (1965), 153-169.

[27] E.Matlis, Injective modules over Noetherian rings, Pacific J.Math. 8 (1958), 511-528.

[28] H.Matsumura, Commutative Algebra, Benjamin,New York,1970.

[29] M.P.Murthy, A note on factorial rings, Arch.Math. 15 (1964), 418-420.

[30] C.Peskine and L.Szpiro, Dimension projective finie et cohomologie locale, Publ.Math.IHES 42, Paris, 1973.

[31] D.Rees, The grade of an ideal or module, Proc.Camb.Phil. Soc. 53 (1957), 28-42.

[32] D.E.Rush, The G-function of MacRae, Preprint.

[33] P.Samuel, On unique factorization domains, Tata Institute of Fundamental Research, Bombay, 1964.

[34] J.-P.Serre, Algèbre Locale - Multiplicités, Lectures Notes in Mathematics 11, Springer-Verlag, Berlin, 1965.

[35] - - , Sur les modules projectifs, Seminaire Dubreil, Paris, 1960-61.

[36] R.Y.Sharp, Finitely generated modules of finite injective dimension over certain Cohen - Macaulay rings, Proc. London Math.Soc. 25 (1972), 303-328.

[37] - - , The Cousin complex and integral extensions of Noetherian domains, Proc.Camb.Phil.Soc. 73 (1973), 407-415.

[38] - - , Gorenstein modules, Math.Zeitschr. 115 (1970), 117-139.

[39] U.Storch, Zur Längenberechnung von Moduln, Arch.Math. 34 (1973), 39-43.

[40] W.V.Vasconcelos, Simple flat extensions, J.Algebra 16 (1970), 105-107.

[41] - - , The commutative rings of global dimension two, Seminar notes, Rutgers University, 1973.

Index